ROUTLEDGE LIBRARY EDITIONS: AGRICULTURE

Volume 16

SETTLING THE DESERT

SETTLING THE DESERT

Edited by
L. BERKOFSKY, D. FAIMAN AND J. GALE

LONDON AND NEW YORK

First published in 1981 by Gordon and Breach Science Publishers Ltd.

This edition first published in 2020
by Routledge
2 Park Square, Milton Park, Abingdon, Oxon OX14 4RN

and by Routledge
52 Vanderbilt Avenue, New York, NY 10017

Routledge is an imprint of the Taylor & Francis Group, an informa business

British Library Cataloguing in Publication Data
A catalogue record for this book is available from the British Library

ISBN: 978-0-367-24917-5 (Set)
ISBN: 978-0-429-32954-8 (Set) (ebk)
ISBN: 978-0-367-25450-6 (Volume 16) (hbk)
ISBN: 978-0-367-25451-3 (Volume 16) (pbk)
ISBN: 978-0-429-28788-6 (Volume 16) (ebk)

Publisher's Note
The publisher has gone to great lengths to ensure the quality of this reprint but points out that some imperfections in the original copies may be apparent.

Disclaimer
The publisher has made every effort to trace copyright holders and would welcome correspondence from those they have been unable to trace.

Settling the Desert

edited by

L. BERKOFSKY
D. FAIMAN
J. GALE

Jacob Blaustein Institute for Desert Research
Ben-Gurion University of the Negev
Sede Boqer, Israel

Gordon and Breach, Science Publishers, Inc.
One Park Avenue
New York, NY 10016

Gordon and Breach Science Publishers Ltd.
42 William IV Street
London WC2N 4DE, England

Gordon & Breach
7-9 rue Emile Dubois
F-75014 Paris

Library of Congress Cataloging in Publication Data

Main entry under title:

Settling the desert.

Includes bibliographies and index.
1. Arid regions agriculture–Congresses. 2. Desert resources development–Congresses. 3. Human ecology–Congresses. 4. Deserts–Research–Israel–Congresses.
I. Berkofsky, Louis. II. Faiman, D. III. Gale, Joseph, 1931-
S612.2.S47 333.73 80-27432
ISBN 0-677-16280-4

Library of Congress catalog card number 80-27432. ISBN 0 677 16280-4.

ישׂשום מדבר וציה ותגל ערבה ותפרח כחבצלת

(Jes. xxxv 1)

Table of Contents

PART TWO: RESOURCE ECONOMICS OF THE DESERT

FOREWORD

One of the most important challenges confronting humanity today is the conversion of deserts and arid zones into productive habitats with a high quality of life. This need is accentuated because an important part of the terrestrial area of the globe is arid, affording its inhabitants only a meagre existence. Indeed, this is where a significant part of the earth's poverty is to be found. World population, according to accepted forecasts, will double itself in approximately a generation and it seems evident that the gap in self-sufficient production of food and material, which is typical of the arid areas, will become much wider. Thus, in the not too distant future, humanity faces a dangerous and intense struggle between the saturated "haves" and the hungry "have-nots."

Therefore, from both human and pragmatic standpoints, the enlightened technological world will have to devote its attention to the development of arid zones as well as to the prevention of desertification, a process which rapidly renders large tracts of land unproductive.

A ray of hope is the belief that with the aid of science and technology, it is in principle possible to increase significantly the bioproduction and therefore the quality of life in many arid zones. Indeed, we sincerely believe that the biblical verse, "...the desert shall rejoice and blossom as the rose..." need not remain a messianic aspiration. Today, scientists can say with confidence that it is possible to exploit the positive conditions for primary plant production which exist in most arid zones, in order to produce a secure and sufficient food supply. It is our conviction that this achievement will be possible by the use of new agro- and bio-technologies based on utilization of natural energies, by selection and development of suitable plant types and by correct usage of water from various sources.

For Israel, the subject of arid zone development is particularly relevant, for it is a very small country of which about half the area is arid or semi-arid. Success in developing ideas and technology for desert settlement is for us, in the long run, a condition for existence. This, then, is the background for the decision to establish a university institute for the study of desert settlement at Sede Boqer. In the institute, diverse research units have been formed, each being dedicated to the study of an individual aspect which relates to the advancement of desert settlement. Free and constant communication is encouraged between these units so as to foster inter-disciplinary research, devoted to the various biological, technological and social concepts involved in arid zone settlement. It is our hope to develop in Sede Boqer an international centre for the study of arid zone settlement - a meeting place for researchers interested in this subject. We would like to teach and be taught, for our own advantage as well as that of mankind.

The works brought together in this book are based on lectures given by the heads of the research units of the Jacob Blaustein Institute for Desert Research on the occasion of its ceremonial opening in November 1978.

Amos Richmond
Professor of Biology
Chairman of the Jacob Blaustein
Institute for Desert Research

PREFACE

Settling the Desert is an attempt to organize those aspects of scientific and sociological research that are the necessary prerequisites for making the desert a comfortable and profitable place for man to inhabit.

Hitherto, the motive for desert research has not been universal, and depending upon one's geographical perspective, different aspects of this research will find differing degrees of importance. Four examples that serve to illustrate this point are the deserts of Australia, the United States, North Africa and Israel. At the risk of gross over-simplification, one can extract four distinct settlement motives from these examples.

In Australia there is a small population with abundant non-arid land, but a need exists to supply comfortable living conditions for people who mine minerals located in the desert.

In the United States there are enormous population centers, but abundant non-arid land areas are still available. Here, massive over-crowding and rising land values have created a not insignificant urge to "get away from it all" and start afresh in the unspoiled desert. In this situation we must add to the need for comfortable living conditions the requisites of energy efficiency and ecological awareness.

Our third example, North Africa, exemplifies the tragic situation that exists in too many of the arid zones of the world: starving populations and no fertile land for food production. Here the research emphasis must be on new and suitable methods of food production.

Finally, we cite the Israeli experience as an example of what lies in store for the whole world: expanding population in a limited fertile area with the desert as the only place to go.

With these motives in mind we have organized the various research topics in this book into three distinct sections. The section on desert agriculture takes a look at what has been done in the past to feed desert populations and goes on to discuss some of the directions from which desert dwellers of the future will derive their nourishment. This section by no means exhausts the many possibilities that exist. In this context one must never forget that we are not talking about the deserts of the moon where self-sufficiency is everything, but rather of a planet with ever-improving transportation networks. Thus, under the heading of desert agriculture we include the bioproduction of commodities that are not necessarily of immediate nutritional value to their producers.

The second section deals with resource exploitation. Here the emphasis is on developing those natural advantages of deserts for use in the desert. Thus for example, mineral exploitation per se finds no place in this book, whereas ecology is placed alongside resource exploitation in importance.

The third and final section of *Settling the Desert* deals with the human factor. In a book that purports to discuss the results of scientific research, this is the most difficult section to include; and yet, man is what the subject is all about. The world population explosion will not adjourn in order to await the discovery of the mathematical equations of anthropology. Architects must continue to use their accumulated knowledge in order to design appropriate structures for desert settlements, whether or not today's scientists are capable of understanding every aspect of their design.

In closing, it should be mentioned that *Settling the Desert* owes its existence to a conference that took place at Ben-Gurion University's Jacob Blaustein Institute for Desert Research. The editors wish to express their deep appreciation to Dr. Shabtay Dover who organized the conference and to the many invited guests whose expert criticism was so important.

Acknowledgement is also made to the following publishers for gracious permission to use copy-righted material. Charles E. Merrill Publishing Co. for Fig. 2 on p. 98 and Fig. 3 on p. 99; The World Meteorological Organization for Fig. 1 on p. 97 and Fig. 7 on p. 103; Plenum Press for Fig. 6 on p. 102; The National Academy of Sciences for Fig. 5 on p. 101 and Fig. 8 on p. 104 and the Electric Power Research Institute for Fig. 7 on p. 141.

Lastly, we particularly wish to express our appreciation to June Burton and Ruth Ma'ayan for typing the manuscript under less than ideal conditions. תושלב"ע

L. Berkofsky
D. Faiman
J. Gale

Sede Boqer, Summer 1980

Part One

DESERT AGRICULTURE PAST AND FUTURE

DESERT AGRICULTURE PAST AND FUTURE

The main feature of deserts, which makes them what they are, is of course lack of water. This is intimately related to high solar radiation intensities, extremes of air temperature and poor and saline soil.

For many centuries man has endeavoured to concentrate the water resources of the desert by water harvesting, and to conserve water by storage in the soil, in artificial lakes and in underground cisterns. Many ancient civilizations prospered in deserts when their water harvesting technology was developed and perished when inadequate maintenance and poor irrigation policy led to silting of the water harvesting systems and salinization of the soil.

Any settlement of the desert and especially desert agriculture must inevitably lead to a disturbance of the natural desert ecosystem. The purpose of rational desert settlement must be to enable man to live comfortably and to earn his living in the desert, while preserving those very features which make deserts attractive, such as open uncluttered spaces and absence of pollution.

In the following chapters, Michael Evenari reviews the history of desert runoff agriculture, Michael Zohary the problems of introducing new plants and renewing the natural unirrigated vegetation of deserts, and Daniel Cohen discusses the special problems of animal husbandry unique to hot desert areas.

Whereas ancient civilizations such as the Nabateans demonstrated the feasibility of agricultural settlements in areas having an annual rainfall of less than 200 mm, it is doubtful whether this type of agriculture would enable a standard of living acceptable to today's farmer. Present research is being directed towards a maximum utilization of the few advantages available in deserts, namely, high radiation (especially during the winter season), runoff water, limited quantities of brackish fossil water and fresh water obtained from solar desalination. How to put these limited resources to maximum use and at the same time to conserve them is discussed by Michael Evenari, Amos Richmond, and Joe Gale.

TWENTY-FIVE YEARS OF RESEARCH ON RUNOFF DESERT AGRICULTURE IN THE MIDDLE EAST

MICHAEL EVENARI

I. INTRODUCTION

Extensive traces of ancient desert agriculture have been found in most deserts of the old world in the Middle East, Arabia and North Africa. The Englishman Palmer[1] was the first European to discover, in 1869, these ancient installations in the Negev and to wonder how it was once possible to cultivate the land in an area which at his time was a barren desert. Did the climate change? Is the presence of ancient agriculture in today's desert an indication that at those times the climate was more humid and the rainfall much higher than today?

We know today that since about 3000-2000 B.C. climatic fluctuations have taken place, as is also typical of today's desert climate, but that there have been no drastic climatic changes. The vegetation and rainfall in the areas in which ancient desert agriculture was carried out were more or less identical to those of today. This is according to palynological and dendroarchaeological research as well as investigations of the changes of the level of the Dead Sea and its salt content.[2,3,4]

Since normal agriculture, using rain where it falls, is not possible in deserts with an average annual rainfall of less than 150-200 mm, and since the ancients could not bring in irrigation water from the outside as we do today, what was their agricultural water source? Many years of study have taught us that it was runoff water, which forms in deserts in comparably large quantities after rains of a certain minimum intensity and duration. This article is too short to go into details of runoff formation in deserts. Such relevant information concerning the mechanics of runoff in the Negev and its relation to rainfall, topography, soil and temperature has been published.[5,6] We will, however, demonstrate the various ancient systems for harnessing the runoff for agriculture by using the Negev Highlands as an example.

II. THE ANCIENT AGRICULTURAL SYSTEMS OF THE NEGEV

a. Terraced Wadis

The terraced wadis represent the most primitive method of harnessing the runoff for agriculture.

Many years ago, when our team (the late Naphtali Tadmor, Leslie Shanon and the author) first began to study the remnants of Negev Highlands ancient desert agriculture in the field, we surveyed the terrain from the air by flying over it in a Piper Cub or a helicopter in early morning or late afternoon. At these hours the long shadows cast by any object on the ground make the object stand out in bold relief and small features of the landscape, which otherwise would have evaded us, are readily visible.

Our attention was first drawn by the many secondary and tertiary wadis which looked like rows of steps on a stepladder (Fig. 1). When investigated on the ground, each "step" turned out to be a terrace with a stone wall built at a right angle to the source of the wadi. The height of the solidly built terrace walls, as seen from the lower terrace, varies according to the depth and slope of the wadi. In extreme cases they may be up to 80 cm high, but they protrude only 10-20 cm above the ground of the upper terrace. It suffices to see a flood streaming down such a wadi and cascading from terrace to terrace in order to understand the agricultural function of the terraced wadis. Each terrace wall retains 10-20 cm of the floodwater which then soaks into the loess soil of the terrace. The loess soil is not very deep and 10-20 cm of water are enough to bring it to its full water holding capacity ("field capacity"). Once this has happened the terrace field is ready for any agricultural use. The terrace walls also serve the additional function of stabilizing structure, preventing erosion.

In 1978 the students of Sede Boqer college reconstructed some terrace walls in a wadi near Sede Boqer and grew onions and other crops on the terraces after a flood had wetted the terrace soil, showing that the ancient method works even today.

b. Farm Units

Even during our first flights we already observed compounds of terraced fields surrounded by stone walls (Fig. 2), mostly in the neighbourhood of the ruins of the ancient cities of Avdat and Shivta. Furrows, each of which terminated

FIGURE 1, Aerial photograph of some terraced wadis in the Negev Highlands.

in a terrace, were visible around each compound. Investigation on the ground revealed that each "furrow" was a channel. We soon realised that these compounds were runoff irrigation systems composed of two functional units: the terraced fields for the cultivation of agricultural crops, and the much larger catchment area from which channels led the runoff onto the terraced fields. The terraced fields and catchment area together formed one indivisible whole, the runoff farm unit. The catchment area for each farm provided the water needed for the crops grown on the terraced fields. In contrast to the terraced wadis which received their water directly from wadi floods, the water source for the terraced fields of the farm units was the water running off the hills and slopes of the catchment area. This water collected in the channels which, by clever use of the local topography, led it onto the terraces. When we surveyed a number of runoff farm units in the field and measured, in each farm unit, the relative size of the terraced fields to the catchment area belonging to it we soon noticed that this proportion, which we called "R" was always of the same order of magnitude and varied only between 1:20 and 1:30. This means that each hectare of the

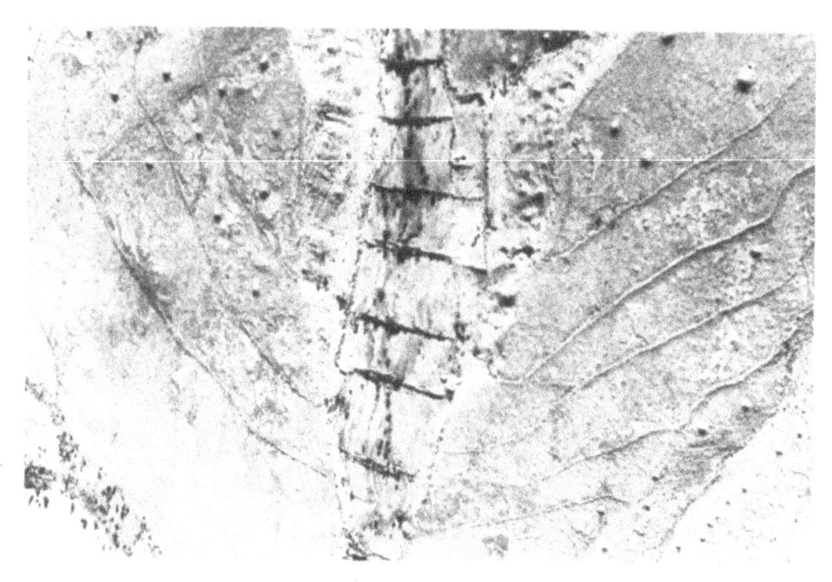

FIGURE 2, Aerial photograph of part of an ancient farm. The terraces, the surrounding wall and the channels leading runoff onto the terraces are clearly visible.

terraced fields received runoff from 20-30 hectares of its own catchment. A purely theoretical calculation will demonstrate the importance of the "R" factor. Our flood measurements over the years have shown that on an average, 15-20 percent of the annual amount of rainwater does not infiltrate into the soil but runs off as floodwater. This very high percentage is explained in the fact that the loess soil, which is characteristic of all areas where runoff farming was practiced, forms a nearly water-impermeable crust after rainfall of a certain minimum intensity and duration. After this crust has been formed most of the additional rain will not infiltrate but run off. The details of this crust and its formation have been described elsewhere.[6,7] If 15-20% of the rainwater runs off and we suppose an annual rainfall of 100 mm and a catchment size of 30 hectares, the cultivated field of a size of 1 hectare will receive 450-600 mm of runoff water, an amount which is sufficient for any type of agriculture.

The question arises what happens to the runoff water once it has reached the terraces. The terrace walls are on an average about 30 cm high and therefore retain 30 cm of water on each terrace, which after a flood soaks into the loess soil of the terraces. The surplus water flows to the next lower terrace through well-built spillways set into each terrace wall. In the beginning of our research we wondered why the ancient farmers made the terrace walls only 30 cm high and did not try to retain more water on the terraces. We understood the reasons when we found out that the loess soil of the terraces was not more than about 3 m deep and that 30 cm of water are just enough to bring the soil to its full water holding capacity. Once the water has infiltrated the soil, the same crust mentioned before is formed on the interface soil-atmosphere. It "seals" the soil against evaporation and the soil loses not more than 8-10 mm of water per growing season by evaporation, and this on an area where about 2.5 m of water evaporate annually from an open air surface. This means that if one good flood has brought enough runoff water onto the cultivated area of a farm unit, there is enough water present for one year's cultivation of crops. The ancient farmers did not "irrigate" with runoff water but stored it in one of the most perfect reservoirs for water, the loess soil, which has a high storage capacity (15-18% field capacity) and loses little of its stored water by evaporation.

Our knowledge of the working of desert runoff farms is not based exclusively on the field survey of the ancient remnants. In 1959/60 we reconstructed two ancient farm units near Avdat and Shivta (Figs. 3a, 3b, 4) where we successfully grow fruit trees (almonds, pistachios, olives, peaches, apricots, grapevines, etc.), field crops (wheat, barley, sunflowers, safflowers, peas, onions, etc.), vegetables (artichokes, asparagus, etc.) and pasture plants[5,8] by using only runoff water as used by the ancient desert farmers.

c. Division Systems

These systems differed in a number of ways from the two described so far. Their cultivated area was considerably larger and was situated along the banks of the large main wadis, their catchment area amounted to many square kilometers. These large catchments produced large amounts of runoff, necessitating much more complex structures. The diversion system near the ruins of the ancient city of Kurnub will serve as a good example. (Fig. 5)

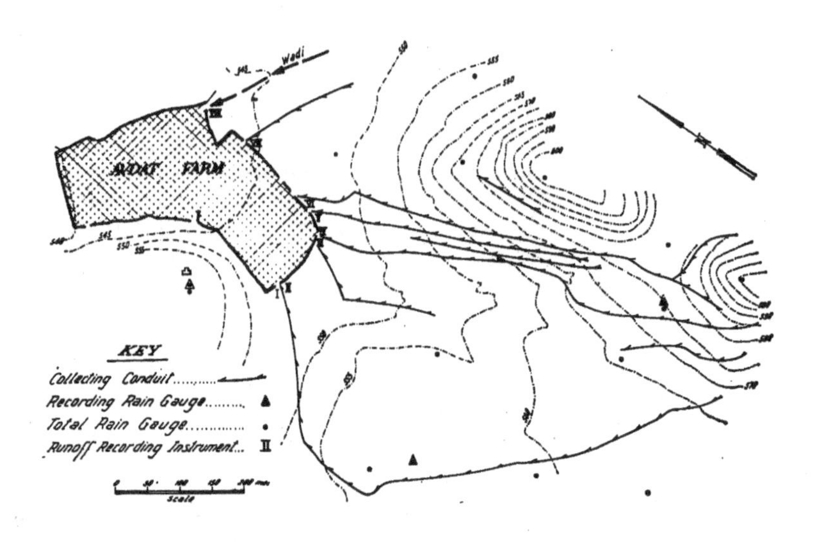

FIGURE 3a, Plan of the reconstructed Avdat farm with its catchment area.

FIGURE 3b, Photograph of the reconstructed Avdat farm. At left and background - fruit trees (almonds, pistachios, peaches, apricots, etc); at right and foreground - sunflowers; at left foreground - onions for seed production.

FIGURE 4, Photograph of Shivta farm with grape vines, olive and carob trees.

Wadi Kurnub is a large wadi in a steep and narrow canyon. After a heavy rainstorm it carries large quantities of floodwater deriving from a catchment area of about 27 square kilometers. At the point where the canyon opens into a broad floodplain, the ancient farmers built a diversion dam into the wadi which diverted part of the floodwater into a large stone-built channel. This diversion channel is 5-9 m wide and about 400 m long. It led the floodwater to a row of broad terraced fields with solidly constructed terrace walls 150-200 m long. Each terrace had at least one large spillway. The cultivated fields covered an area of 10-12 hectares.

d. Microcatchments

Many years ago Joseph Waitz, the late director of the Jewish National Fund, visited the steppe and desert region of South Tunisia and saw that the peasants there success-

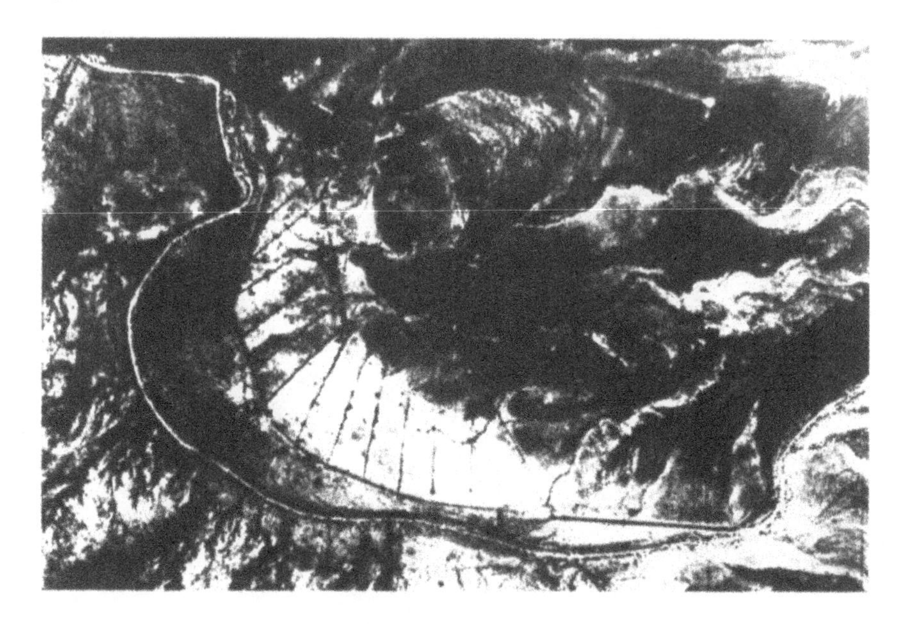

FIGURE 5, Aerial photograph of the diversion system at Kurnub. Wadi Kurnub flows from right lower corner into the floodplain. A diversion channel (straight black line) leads part of the wadi's floodwater onto the large terraced area.

fully grew olive and palm trees in "microcatchments" by using runoff as a water source (Fig. 6). The basic principle of this method is very simple. Instead of collecting runoff from a large catchment and leading it into comparatively large fields, a single tree is planted in its own small catchment which provides the runoff water for the one tree. When our agricultural advisor at that time, in 1961, Joel de Malach from Kibbutz Revivim, told us the story of Waitz and the Tunisian microcatchments it appealed to us because our runoff experiments had already taught us that under arid conditions the smaller the catchment the higher the relative water yield per unit surface. We therefore constructed a number of microcatchments by dividing a plain area near our Avdat farm into squares surrounding each square with a low border check 15-20 cm high; we dug a square basin at the lowest point of each microcatchment and planted one fruit tree into each basin (Fig. 7, 8). The microcatchments were a big success. We found that under the rainfall conditions of Avdat the optimal size of micro-

FIGURE 6, Photograph of an olive tree in its microcatchment near Matmata, S. Tunisia. (Photo: Udo Nessler)

catchments for fruit trees was 200-250 square meters. As predicted, the microcatchments produced more runoff per unit surface, rains which do not lead to runoff production on large catchments produced runoff on microcatchments, and, moreover, microcatchments are cheaply constructed and maintained.

When we first used microcatchments we thought that they were a modern invention. Today we know that like the terraced wadis, farm units and diversion systems, microcatchments are of ancient origin. This is why we deal with them here in the framework of a paper on ancient desert agriculture. We know that the microcatchment system in South Tunisia was introduced by the Phoenicians and has been uninterrruptedly used there since. It may be that the ancient farmers in the Negev used microcatchments. If they did, they left no traces which is easily understandable because of the fragility of the structure.

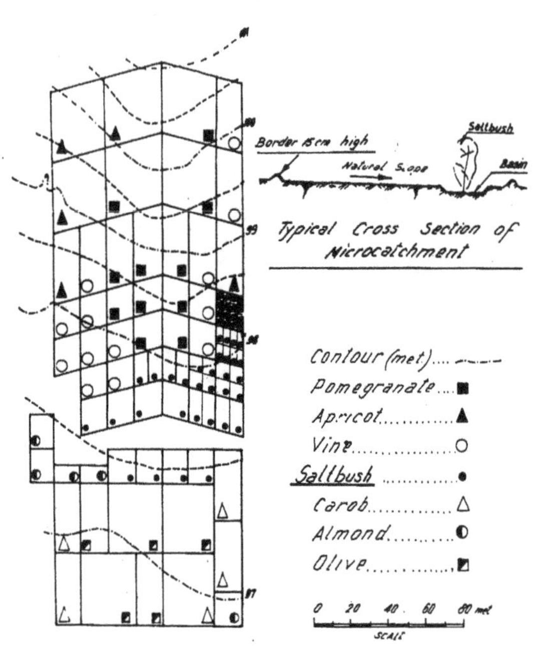

FIGURE 7, Plan and cross-section of microcatchments at Avdat.

FIGURE 8, Almond trees planted in microcatchments in Wadi Mashash.

III. ORIGIN AND HISTORY OF ANCIENT RUNOFF AGRICULTURE

a. In the Negev

The Negev was inhabited during the various stone ages at least from the Middle Paleolith on. But stone age man was a hunter and food gatherer and did no farming. The people of the Middle Bronze I period (MBI, ca. 2100-1900 B.C.) also lived in the Negev Highlands. This period coincides with the time of the Patriarchs when Abraham roamed through the Negev with his family and flocks (Gen. 20:1). The MBI people may have been the first runoff farmers in the Negev. Agricultural implements were found in their villages in excavations carried out by Kochavi,[9] but we did not find agricultural fields which could be dated to this period.

From the 10th century B.C. (Israelite period II-III or Iron Age II) to the 6th century B.C. our forefathers, the Israelites, occupied the Negev and lived there in numerous villages, fortresses and farmsteads. The late Aharoni and our team [10,11] excavated an Israelite village and farm and found many cisterns with their runoff channels dating back to this time (Fig. 9). The Bible gives us additional evidence in reporting from King Uzzia (ca. 782-740 B.C.): "Also he built towers in the desert and digged many wells.. for he loved husbandry" (II Chronicles 26:10). There is no doubt that the Negev Israelites practiced runoff farming in terraced wadis and farm units.

The next runoff farmers were the Nabataeans who made their appearance in the Negev around 300 B.C. They were traders originating somewhere in South Arabia and monopolized the caravan trade bringing spices, perfumes and silk from South Arabia and the Far East to the North. They built the Negev cities (Avdat, Shivta, Kurnub, Khalutza, Nitzana, Rehovoth haNegev) and constructed their numerous agricultural runoff installations and runoff cisterns around these cities. The Romans occupied the Nabataean empire peacefully in 106 A.D. The Nabataean population remained under the Roman and later the Byzantine rule, and during the Byzantine period (392 A.D. - 637/641 A.D.) runoff farming reached its peak in the Negev. In 637/641 A.D. the Arabs conquered the Negev. During the following hundred years the population left the cities and abandoned runoff agriculture. By the 8th century the Negev cities were empty shells and the farms lay waste. The Bedouin

FIGURE 9, An Israelite cistern near Mizpe Ramon, full of water. One of the channels leading runoff water from the hillside through a spout (right centre) can be clearly seen.

were now masters of the Negev. Nothing more vividly describes the deterioration and destruction of the Negev than the passage with which Palmer concludes the report of his trip through the Negev in 1870: "Long ages ago the word of God had declared that the land of the Canaanites and Ammorites should become a desolate waste: that the cities of the south shall be shut up and none shall open them" (Jer. 13:19), and here around us, we saw the literal fulfillment of the dreadful curse. Walls of solid masonry, fields and gardens compassed around with goodly walls, every sign of human industry was there, but only the empty names and stone skeletons of civilization remained to tell of what the country once had been. There stood the ancient towns, still called by their ancient names, but not a living thing was to be seen, save when a lizard glided over the crumbling walls, or screech owls flitted through the lonely streets." In this state of desolation the Negev remained for the next 1200 years until we revived its runoff agriculture.

b. Runoff Agriculture in Other Regions

In the preceding chapter we dealt with the history of runoff agriculture in the Negev. But was it only practiced there? Who invented it? Recent excavations in Jawa (Jordan) have shown that around 3000 B.C. people there already used runoff which was either diverted from wadis or collected from hillsides in channels for filling reservoirs and for cultivation of fields.[12,13] This early use of runoff is not too astonishing since cisterns filled by channels with runoff water already existed in Canaanite cities of the Chalcolithic period (ca. 5th-4th millenium B.C.) as proven by excavations.[14] We may be permitted here to make a speculation. K.Kenyon[15] the famous excavator of ancient Jericho opines that in the Pre-Pottery Neolithic period (ca. 7000 B.C.) the people of Jericho already irrigated their fields with channels using the water of the local spring. Is it too far fetched to believe that the same people living in the arid environment of Jericho who knew how to build channels and divert spring water, used the floods of Wadi Kelt for flood water irrigation? If this speculation is right, runoff farming is even older than the Chalcolithic period.

The most monumental remains of ancient runoff agriculture are found in South Arabia near the village of Marib (North Yemen) which was once the capital city of the mighty Sabaean Kingdom. These are the remains of the famous dam of Marib which was part of an ancient water spreading system. It was an enormous construction with solidly built, gigantic sluice gates at its Northern and Southern ends (Fig. 10). The dam was built across the large Wadi Dhana, made of earth faced with stones and was 700m long, 15m high and 80 m wide at the base. It could retain 60-400 million cubic meters of Wadi Dhana's flood water. The dam was not built to store flood water but to raise it to such a level that it could flow through the two sluice gates. Each gate was connected to a large channel (Fig. 11) which led the water through secondary and tertiary sluice gates into hundreds of small channels which spread it over two field complexes lying along the Northern and Southern flanks of the Wadi (the "two gardens" mentioned by the Koran, see below). It has been calculated that about 20,000 ha could be irrigated by the flood waters from the dam and that about 300,000 people could be fed by the agricultural products of this runoff installation. The dam was dated to ca. 750 B.C., but according to Bowen[16], one of its explor-

FIGURE 10, Ruins of the southern sluice gate of the dam of Marib. (Photo: Udo Nessler)

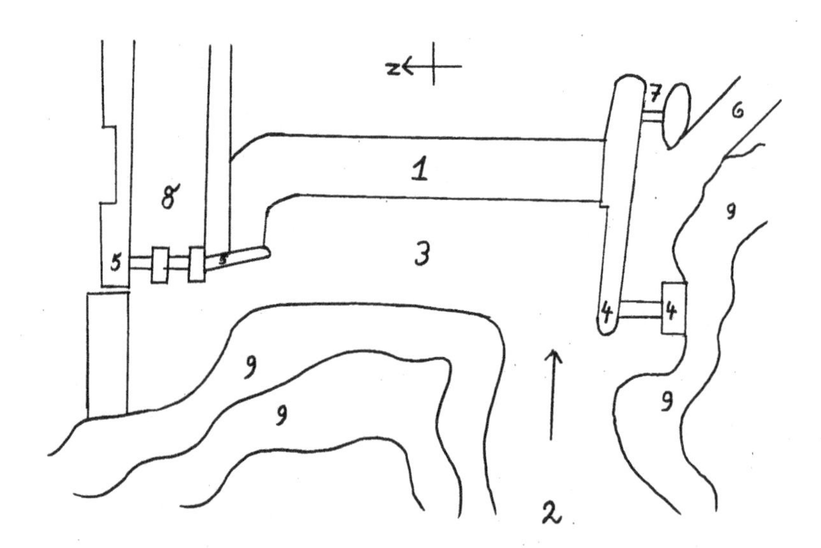

FIGURE 11, Plan of ancient dam of Marib. 1-dam; 2-Wadi Dhana leading floodwater toward dam; 3-reservoir filled by floodwater; 4,5-southern and northern sluice gates; 6,7,8-channels leading water to secondary and tertiary sluice gates; 9-rocks.

ers, "it is reasonable to suppose that it may even antedate this". The dam functioned (with many repairs) for at least 1300 years and the Koran reports its destruction by a colossal flood: "A sign there was to Saba, in their dwelling places, two gardens, the one on the right hand, and the one to the left. 'Eat ye of your Lord's supplies, and give thanks to him: Goodly is the country, and gracious is the Lord'. But they turned aside: so we sent them the flood of Iram and we changed their gardens into two gardens of bitter fruit and tamarisk and some few jujube trees" (34.Sura: 16-17). This destruction took place in ca. 575 A.D. There are now plans to reconstruct the dam more or less in its ancient dimensions. 1400 years after its destruction the Marib dam will rise again and will again produce runoff water and food for many people.

The dam of Marib was not the only ancient runoff installation in South Arabia. In other parts of that region, for example in Hadramauth and Wadi Beihan, extensive traces of ancient desert agriculture have been found by a number of explorers. They are the exact counterpart of what we described for the Negev. Wadi Beihan is located in the ancient Kingdom of Qataban. Bowen[16] states that the ancient Qatabanian runoff agriculture "cannot be older than the second millenium B.C.", i.e. it is very old. According to recent visitors in the area the peasants of today's village of Marib irrigate about 1700 ha with flood water. We can perhaps assume that Marib is one of the few places on earth where runoff agriculture has been practiced continuously since ancient times.

The most impressive ruins of beautifully built large temples, cities and the dam of Marib, the many stone hewn inscriptions in the graceful ancient script bear witness to the importance, wealth and might of the ancient Kingdoms of Qataban and Sheba.

The Bible corroborates this archaeological evidence when, in I Kings 10:1-10, it tells us of the visit of the Queen of Sheba to King Solomon which must have taken place in the 10th century B.C. "And when the Queen of Sheba heard of the fame of Solomon...she came to Jerusalem with a very great train, with camels that bore spices and gold very much and precious stones ... And she gave the King a hundred and twenty talents of gold, and of spices very great store, and precious stones; there came no more such abundance of spices as these which the Queen of Sheba gave to King Solomon".

This story is proof of contacts between the Kingdom of Sheba and that of Solomon as early as the 10th century B.C. At that time runoff agriculture was already highly developed in South Arabia. Could it be that this Sabaean knowledge was transmitted to Solomon or to his successors?

Another region with innumerable remains of ancient runoff agriculture is the desert fringes of North Africa especially in Southern Algeria and Tunisia. The early French exploreers of the region were the first to observe the many traces of agriculture in a region which today is a desert without any cultivation. Some of them already recognized that the fertility of these areas in ancient times was not due to a better climate but to the ingenious use of runoff waters. Some, like Carton,[17] even thought that these ancient methods could be usefully applied in modern times to restore the fertility of the once flourishing region. One realises how large the once runoff cultivated area was by looking at the many air photographs taken by Barradez, a colonel in the French Air Force who discovered in Southern Algeria the traces of "Fossatum Africae", a Roman defensive frontier trench extending over hundreds of kilometers. All along the "fossatum" are found many fortresses, fortified towers, terraced wadis, diversion systems, large runoff fields, runoff cisterns and villages, all this in an area which today is completely barren. As stated by Barradez:[18] "The contrast between the traces of what once existed and the aspect of what remains today is dreadful. The region is terribly devastated ...". The "fossatum" was started by the Roman emperor Hadrian (117-138 A.D.) and garrisoned by "limitanei", i.e. soldier-farmers. Historically these could be regarded as the forerunners of Israel's soldier-farmers, the Nahal, who settle in newly founded frontier-Kibbutzim. The Roman "limitanei" cultivated the area around the "fossatum" by using runoff farming methods. But his was not restricted to Southern Algeria and Tunisia. Wherever the Romans ruled in North Africa and wherever the rainfall was insufficient and no other water source was available there was extensive runoff farming.[17,19]

It may be that the Romans were not the first to practice runoff agriculture in North Africa. When Carthage was founded by the Phoenicians (ca. 814 B.C.) they introduced the culture of grape vines and olives into North Africa and a Carthaginean agronomist by the name of Mago wrote a 28 volume treatise, which is mentioned by Varro (116-27 B.C.) and Columella (1st century A.D.)[20,21] but is lost to us, on agriculture and agricultural economy. It must have been a very important document since the Roman senate decreed its translation from the Punic language into Latin[20] and Columella called Mago "the father of husbandry". It is not too far fetched to take the possibility into account that the Phoenicians had already introduced runoff farming into Tunisia. One sentence of Diodorus Siculus (1st century B.C.) offers some proof of this. He describes the Carthaginean agriculture as follows:[22] "The intervening country through which it was necessary for them (the Sicilian soldiers) to march was divided into gardens and plantations of every kind, since many streams of water were led in small channels and irrigated every part". This invasion of Carthage happened in the 4th century B.C. Tixeront,[23] one of the French authors trying to trace the history of agriculture in North Africa, goes even one step further back. He claims that long before the founding of Carthage, sometime in the second millenium B.C., people from Canaan migrated to Tunisia by sea or overland and that these Canaanites implanted the principles of runoff agriculture into Tunisia, which naturally presupposes that runoff agriculture was known in Canaan already in the second millenium B.C. (or earlier, see p. 17). We may accept Tixeront's theory or not, one fact is certain: runoff agriculture in South Tunisia is ancient and has been successfully practiced there up to our times (see above).

IV. MODERN RUNOFF AGRICULTURE

After our reconstructed runoff farms in Avdat and Shivta had shown that by using the ancient methods fruit trees, field crops vegetables and pasture plants could be grown successfully (see above), we recognised the potential agricultural value of runoff farming for today's arid and semi-arid areas. We did not want to jump to premature conclusions and decided that it was important to verify the positive results obtained in Avdat and Shivta in a much larger farm which would serve as a pilot plant between the two experimental farms Avdat and Shivta and the practical application of our knowledge in commercial or semi-commercial ventures. We erected the pilot farm in Wadi Mashash, about 20 km south of Beersheba (Fig. 12). We planted a few thousand olive, almond (Fig. 13) and pistachio trees in microcatchments and limans (check dams), prepared an improved pasture by planting and sowing pasture plants which in the Avdat experiment had proved their value and started

FIGURE 12, Flood in Wadi Mashash.

FIGURE 13, Almond trees planted in limans in Wadi Mashash.

grazing with sheep (Fig. 14). Of all the fruit trees tried

FIGURE 14, Sheep grazing in summer on improved pasture in Wadi Mashash. Note the bare non-improved pasture in the background.

out, the pistachio and to a lesser degree, the almond seem, under Israeli conditions, to be the most promising candidates for commercial plantations in the Negev because both trees are well adapted to the Negev's climatic conditions and yield well when receiving only runoff as their water source. To prove this point on a large scale a new runoff plot of 5 ha was prepared near Avdat, with the help of Prof. Spiegel-Roy from the Volcani Institute of Agriculture, on which nearly 2000 trees were planted. We also inaugurated a model desert park on an area of 1.3 ha near Avdat, where we planted about 6000 ornamental trees, shrubs and herbs which receive their water only from runoff. The aim of this park is to show that shady green park oases for recreation can be created in desert areas without wasting precious and costly irrigation water.

With the establishment of the Wadi Mashash farm many agencies working in developing countries became interested in our work. With their help a number of developing countries have started runoff farming in their arid areas. In Afghanistan (Province of Khost, Fig. 15) 70,000 ha are

FIGURE 15, Flood irrigated area in Khost (Afghanistan). (Photo: Udo Nessler)

farmed in this way and runoff installations have been established in India, Australia, Botswana, Niger (Fig. 16), Upper Volta (Ougadougou, Titao), Kenya (Lokitarri, Lokori, Maralal), Argentina, Mexico and in some Arab countries.

FIGURE 16, Terraced wadi in Tchirozerine (Niger, Sahel zone), developed as pasture. (Photo: Udo Nessler)

V. REFERENCES

1. E.H. Palmer, The Desert of the Exodus, (Cambridge, 2 Vol., 1871).

2. N. Lipschitz and Y. Waisel, Isr. Explor. J., 23, 30 (1973).

3. A.E. Marks (ed.), Prehistory and Paleo-Environment in the Central Negev, Israel (2 Vol., SMU Press, Dallas, 1976, 1977).

4. D. Neev and O.K. Emery, Bull. Geol. Survey Israel. 41,1 (1967).

5. M. Evenari, L. Shanan and N. Tadmor, The Negev: The Challenge of a Desert (Harvard Univ. Press 1971).

6. L. Shanan, Ph.D. Thesis, Hebrew Univ. Jerusalem (1975).

7. D. Hillel, Agric. Res. Sta. Beit-Dagan Bull. 63 (1959).

8. M. Evenari, U. Nessler, A. Rogel and O. Schenk, Fields and Pastures in Deserts. ed. E. Roether (Darmstadt 1974).

9. M. Kochavi, Ph.D. Thesis, Hebrew Univ, Jerusalem (1967).

10. Y. Aharoni, M. Evenari, L. Shanan and N.H. Tadmor, "The Ancient Agriculture of the Negev" V. Isr. Expl. J., 10, 23 and 97 (1960).

11. M. Evenari, Y. Aharoni, L. Shanan and N.H. Tadmor, Isr. Expl. J., 8, 231 (1958).

12. S. W. Helms, Levant, 8, 1 (1976).

13. N. Roberts, Proc. Seminar Arab. Stud. London, 7, 134 (1977).

14. M. Dothan, Isr. Expl. J., 7, 217 (1957).

15. K, Kenyon, Archaeology in the Holy Land (Methuen, London 1965).

16. R.L.B. Bowen, In: Archaeological Discoveries in South Arabia, (John Hopkins Press, Baltimore 1958), p. 43.

17. Carton, Bull. Archéol. (1888).

18. J. Barradez, Fossatum Africae. (Ed.Arts et Métiers, Paris, 1948).

19. J. Despois, L'Afrique du Nord (Presse Univers. de France Payot, Paris 1964).

20. L.J.M. Columella, De Re Rustica 1, 33 and 35, Loeb Class. Libr. (Harvard Univ. Press 1948).

21. M. T. Varro, De Agri Cultura, 1, 165 Loeb Class. Libr. (Harvard Univ. Press 1964).

22. Diodorus Siculus, The Library of History, 20,161. Loeb Classical Library (1962).

23. J. Tixeront, Karthago, 10, 3 (1959).

PLANT INTRODUCTION IN THE DESERTS OF ISRAEL AND ITS THEORETICAL BASIS

M. ZOHARY

I. PROBLEMS AND AIMS

Quite a number of countries have both sown areas and deserts, but only few are worried about whether their deserts possess enough arable land to satisfy their agricultural needs, or possess large water bodies to enable the expansion of the arable areas. Israel is largely worried about its deserts because they constitute half, or more, of the country's area. Therefore, much effort has been spent in the last decades in desert research with the aim of finding means and methods of utilization of this enigmatic land (Map 1)[1]. It seems that, despite these efforts, a comprehensive concept for the development of the agro-ecological potentialities of our deserts is still lacking. The following pages are aimed at providing some suggestions in this direction and, above all, to address one aspect of it, namely: to elucidate the possibility of utilization, through introduction of plants into part of the desert - classified here under "the green hammada zone"[2] which bears the bulk of the desert's phytomass and which today cannot be utilized by man or herd. For this purpose we wish to briefly review the deserts of Israel in their entirety.

II. THE DESERTS OF ISRAEL AND THEIR ECOLOGICAL DIVERSITY

Israel's deserts are located at the meeting place of three world deserts, namely: the Sahara, Central Asian and the Sudano-Sindian. They comprise the Judean Desert, the Negev and the Arava Valley. They are extremely heterogeneous in their morphology, soils and climate, and accordingly, richly differentiated in their flora and vegetation. A few salient traits of this diversity may be mentioned here:

a) The Judean Desert (Fig. 1) is a stony plateau, gently sloping eastwards, but abruptly falling down to the Dead Sea into which a few lateral wadis empty. It is a rain-shadow

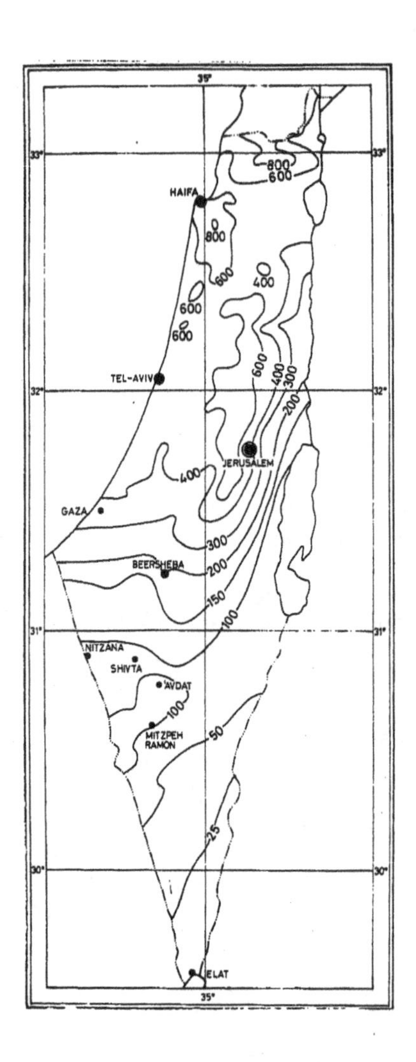

MAP 1, Mean annual rainfall in Israel (Atlas of Israel)[1]

desert, built up of soft, calcareous rocks alternating with rugged hard limestone ridges.

b) The Negev is exceedingly differentiated morphologically. In the north and west it is marked by flat areas of sands, sandy loess and loess (Figs. 2-6). This flatland is surrounded in the north and east by a series of Eocene soft hills covered with seroziems (Fig. 7) or on-blown loess. Toward the central Negev there is a rugged plateau rising up to 1,000 m and made up mainly of Cenomanian and Turonian hard limestone rocks (Fig. 8). This plateau is cut by a series of deep wadis. South of Mt. Ramon, the plateau drops suddenly into an enormous flat area out of which many hills and small ridges emerge. This eroded plain is cut by numerous wadis and their tributaries which abound with vegetation, including arboreal (Fig. 9).

The Arava Valley is part of the Syro-Eritrean Great Rift Valley. It displays six main landforms: the saline marshes south of the Dead Sea, Yotvata and Eilat (Figs. 10, 11), the sterile block of the Lissan marl, south of the Dead Sea, the sand flats and dunes derived from Nubian sandstone (Fig. 13), the alluvial fans at the outlet of the wadis (Fig. 12) and igneous rock massive and its derivatives in the Eilat region (Fig. 14).

This diversity of landforms has brought about innumerable habitats, each with its own flora and vegetation.

FIGURE 1, Judean Desert. Marly slopes with Chenoleetum arabicae.

FIGURE 2, Loess soil in northern Negev. Broken loess layer, partly under cultivation.

FIGURE 3, Cultivated loess plain west of Beer Sheva with *Aellenia hierochuntica* as a (tumble) weed.

FIGURE 4, Virgin (not cultivated) loess cover with *Artemisietum herbae albae* in Edom (Jordan).

FIGURE 5, Sandy landscape of the Sinai coastal area (west of El-Arish) with Stipagrostis scoparia and date palms.

FIGURE 6, Sandy area of the northeastern Negev with Alabasetum articulatae (Asphodelus microcarpus in flower).

FIGURE 7, Hills north of Beer Sheva with gray soil (seroziem) covered with Artemisietum herbae albae.

c) Of the climatic features of the desert, mention will be made of a few data concerning temperature and rainfall. Temperature is, in our opinion, best expressed by the mean maxima for the hottest month and the mean minima for the coldest (in °C), showing for Beer Sheva: 26.3 - 11; Mitzpe Ramon: 24.7 - 9.3; and Eilat: 38.8 - 15.9. For further data see Atlas of Israel. From these and other data not reported here, one may readily conclude that there is no single point in the desert where prevailing temperatures exclude, or even obstruct, plant life. Moreover, these drastic differences in temperature cause great variation in the flora and vegetation of the desert. So, for instance, the plateau of the central Negev harbors quite a number of "cold" Irano-Turanian plants, while a few miles further east there is a savanna-like vegetation of Sudanian origin. In general, it could be said that temperatures alone play only a minor part in plant life when compared with rainfall.

Rainfall is the most decisive factor in the life and death of plants, and in Israel's deserts there are quite a number of places which are altogether plantless because of insufficient rain. The decrease in the yearly amount of rainfall from north to south is much more abrupt in the desert than in the sown area. The following remarks may be added here as to the relationship between the rain factor and plant life:

1) Rainfall efficiency depends largely on soil quality, on surface configuration and on the aspect of the slope.

FIGURE 8, "Green hammada," central Negev near Sede Boqer. Mainly Zygophylletum dumosi.

2) Though average data on climatic factors are of little ecological value, rainfall averages have much to say in plant life, perhaps because rainfall is cumulative. In other words, the amount of rain in an excessively rainy year is felt not only in the year concerned, but also in one or more successive years.

3) Because the efficiency of rainfall varies so markedly according to soil properties, one is unable to assess that amount which is the biological threshold for higher plants. So for instance, the hills of the Negev plateau and the valleys between them are relatively densely vegetated, despite the fact that the local amount of annual rainfall does not exceed 80 mm. The truth is that not only do the plants of the valley enjoy runoff water in addition to local rainfall, but the hill vegetation also enjoys additional moisture, due to the fact that the coarse stones covering the hill form micro-dams, preventing the downward flow of the rainwater. Moreover, considering the fact that the root systems of many dominant plants are made up of a network of horizontally and shallowly spreading moisture-absorbing branches, one will not be astonished at the occurrence of such an enormous phytomass on these hillsides within the 80 mm isohyet. This shows how careful one should be in fixing the relationship between isohyets and plant life.

FIGURE 9, Southern Negev. Plantless mountain and hill ridges protruding from vast reg plains cut by shallow wadis, inhabited by tamarisk and retam bushes.

FIGURE 10, Sebha in the northern foreshore of the Red Sea. Large saline mainly with Suaedetum monoicae.

FIGURE 11, Saline flat in Yotvata (Arava Valley) with Desmostachya bipinnata association. Wild date palms on the horizon.

FIGURE 12, Arava Valley. Colluvial fans with Acacietum raddianae.

FIGURE 13, Sandy flats (sand derived from Nubian sandstone) with Haloxylon persicum (Arava Valley).

FIGURE 14, Alluvial plantless flats at the foot of igneous rock mountains. Arava Valley (Um Rashrash in 1940, Eilat of today).

MAP 2, Outlines of an agro-ecological vegetation map of Israel's deserts.

d) The soil factor, although its ecological importance is much less than that of rainfall, plays a remarkable part in the efficacy of moisture absorption, retention and mobility. The physical properties of the soil are far more important than the chemical ones. Of the many soil varieties recognized by soil scientists, the following are geobotanically the most outstanding: sandflats and sand dunes (Figs. 5, 6 and 12), loess, sandy loess (Figs. 2-4), sieroziems (Fig. 7), hammada (Fig. 15), regs (Fig. 16), and salines (Figs. 10, 11 and Map 2).

LEGEND TO AGRO-ECOLOGICAL DISTRICTS

A. Flatlands (arable)
B. Green hammadas with intermountain valleys (runoff farming and pasture land)
C. Streambed lands
D. Regs and black (sterile) hammadas
E. Arava valley (alluvial fans, sands, marshes; tropical)

LEGEND TO VEGETATION MAP

1. Mediterranean batha and garigue on kurkar hills of the coastal plain; mainly littoral variants of *Sarcopoterietum spinosi*, *Coridothymetum capitati*, *Calycoteometum villosae*, *Cistetum* (in the northern part). Comprises also *Helianthemetum elliptici* (in the southern part).

2. Mediterranean batha with remnants of evergreen maquis and forests of the *Quercus calliprinos* - *Pistacia palaestina* assoc.

3. Climax district of evergreen maquis and forest, destroyed and reoccupied by typical Mediterranean batha and garigue vegetation, e.g., *Sarcopoterietum spinosi*, *Calycotometum villosae*, *Cistetum salvifolii*. Segetal vegetation consists of the *Carthamus tenuis* - *Ononis leiosperma* assoc.

4. Sand and sandy hammada deserts in Arava Valley and Edom, dominated by the *Hammadetum lalicornici*, the *Haloxyletum persici* and *Acacietum raddianae*.

5. Barren hammada deserts of central and southwestern Negev. Vegetation mainly in depressions and wadi beds, consisting of the *Anabasidion articulatae* alliance and the *Acacietum tortilis* assoc. in wadi beds.

6. Irano-Turanian dwarf shrub steppes on western escarpments of Transjordan. Vegetation not adequately known, probably consisting of the *Artemisietalia herbae albae* order.

7. Steppes of *Artemisia monosperma* and annual grasses on sandy loess soils of western Negev.

8. Saharo-Arabian hammada deserts consisting mainly of the following alliances: *Anabasidion articulatae*, *Zygophyllion dumosi*, *Chenoleion arabicae*, *Suaedion asphalticae*. (Judean Desert).

9. Mosaics of *Zygophyllion dumosi* on hillside hammadas and *Hammadion scopariae* on loess plains.

10. Saharo-Arabian hammada deserts of central Negev, consisting mainly of the *Zygophyllion dumosi* alliance.

11. Mobile and semi-mobile sand dune vegetation of coastal Negev, consisting mainly of the *Artemision monospermae*.

12. Salines of Arava Valley, lower Jordan Valley inhabited by the *Arthrocnemum glaucum-Tamarix tetragyna* assoc., *Suaedetum monoicae, Atriplicetum halimi, Salsoletum tetrandrae, Nitrarietum retusae*.

13. Climax district of Irano-Turanian dwarf shrub steppes, mainly of the *Artemisietalia herbae-albae* order.

14. Irano-Turanian steppes on loess soil; primary vegetation almost entirely destroyed. Segetal vegetation: mainly *Achilleetum santolinae, et al.*

15. Sand dune vegetation of interior Negev, mainly of the *Anabasidion articulatae* alliance.

16. Enclaves of tropical vegetation, mainly *Zizyphus spina-christi-Balanites aegyptiaca* assoc. and *Acacietum raddianae*.

17. Alluvial plains: climax vegetation destroyed and mostly unknown. In some parts remnants of subtropical *Hyparrhenia* savannah. Segetal vegetation consists mainly of the *Prosopidion farctae* alliance.

FIGURE 15, "Black sterile hammada" in Transjordan.

FIGURE 16, Reg ground (desert pavement) in central Sinai. Altogether sterile, except for lichens and algae.

The flora of Israel's deserts are relatively rich. No less than 1200 species have been recorded so far. There are only a few endemics here because the local deserts are direct continuations of those in the adjacent countries. The phytogeographical nature of the flora is rather complex because it comprises elements of the Irano-Turanian, Saharo-Arabian, Mediterranean and Sudano-Sambezian regions. Accordingly, the deserts have been divided into various phytogeographical territories.[6,7] This phytogeographical diversity is most promising with regard to introduction.

The vegetation is also rather complex. Although most of Israel's vegetation orders have their representatives in the deserts, the bulk of the desert vegetation belongs to the classess of Anabasetea and Artemisietea, which are exceedingly rich in plant communities.

III. THE AGRO-ECOLOGICAL SUBDIVISION OF THE DESERTS

After reviewing the Israeli deserts from different aspects, we wish to devote the following pages to the main subject matter, namely, the utilization of the desert from the viewpoint of applied geobotany.

Since there is as yet no comprehensive mapping of the agro-ecological potentials of Israel's deserts on hand, the following suggestion as to the agro-ecological subdivision of the desert will perhaps provide the interested bodies with a clearer outlook on the land inventory and its ecological potentialities. The divisions suggested are based on first-hand knowledge and reconnaissance:

a. The soft flatlands of the northern and northeastern Negev (Figs. 2-6). This includes the extensive area of the loess and its derivatives, the loess-sand area, and the sand flats and dunes. This is the most fertile land zone of the Negev. Most of this area has been cultivated for centuries, and part has even been able to produce summer crops as well (dura and melons). Of course, without irrigation this area yields market crops only in rainy years.

It is this land zone, consisting mainly of the three above-mentioned soil units and marked by their appropriate flora and vegetation, that constitutes the most promising land resource of the Negev, and it is this part of the Negev which should be settled first and receive first priority in water allocation. In any event, it is not the right region for the introduction of Atriplex or Simmondsia.

b. The runoff lowlands. This is the area of runoff farming (Figs. 2, 8). It comprises a large part of the stony and hilly regions of the northern and central desert, serving catchment zones for the inter-mountain or intercoline valleys. The ancient methods of establishing runoff farms so thoroughly studied and modernized by Evenari et al.[3] will play a significant role in the future farming of the desert. The extension of this type of agriculture will, for lack of alternatives, remain the only way to make use of the runoff area, although much hydrotechnical skill and investment will be required to build up and maintain these runoff farms.

c. The dry stream bed zone. The central and especially southern Negev is cut by an enormous net of rather shallow and broad wadis, extremely rich in tributaries. These dry wadi beds are the only sites supporting vegetation, sometimes even arboreal. The sum total of this furrowed land mass is estimated to comprise no less than a million dunams (100,000 hectares) which through appropriate hydro-technical planning could somehow be agriculturally utilized. It is a question of harnessing floodwater and of water collection, which are not easily carried out. But as it is, it forms an ecological zone of its own and displays peculiar potentialities.

d. The green hammada. The most salient feature in the desert landscape are the ridges and plateaux built up of soft and hard limestone, covered with coarse stones. This is what pedologists and plant ecologists call hammada (true hammadas).[2] These hammadas display in the north a continuous, though spaced, cover of vegetation (Fig. 8) belonging to two major vegetation classes,[6,8] the Irano-Turanian Artemisietea and the Saharo-Arabian Anabasitea. They are called here green hammada to distinguish them from sterile (black) hammada (Fig. 14) of the southern Negev. These extensive ridges and hillsides not only feed their adjacent valleys with runoff water, but sustain a rather interesting plant world of their own. Occupying over thirty percent of the surface of Israel's deserts, this is the mainstay for Israel's desert settlement. It is this zone which concerns those who are considering the future of Israel's deserts. For the present author it constitutes the main topic of the present article and the focus of the introduction problem. As it is at present, it harbours the main phytomass of the desert, amounting to millions of tons of plant material and yielding an average annual production of about half a million tons of green matter. This green material, unusable by the grazing herd, represents a marked loss of energy.

e. The regs and the black hammada (figs. 14, 15). About two million dunams of Israel's desert lie bare and plantless and are doomed to desolation forever. These are the regs, the immense black flats, armored with gravel and rubble forming a compact pavement. Beneath is a layer of deep powdery soil, rich in gypsum and soluble salt. Here and there a dry river-bed cuts these plains, or hills and hillridges emerge from them. Both pavement and the hills are desolate of plants because they are situated in the driest part of the Negev and because the few drops of rain which they occasionally do receive dry up on the surface of the ground without being able to penetrate the soil. These lands are not included in any framework of utilization projects. In the long-run they may perhaps be looked on as the lungs of Israel, destined for resorts and tourism.

Apart from these five major units, which can hardly be delineated exactly, some smaller sites with interesting features of their own are included in the desert of Israel. These are the salines, river banks and springs. They are botanically of great significance, but cannot be dealt with here as landforms.

IV. INTRODUCTION AND ITS THEORETICAL BASIS

After the lengthy survey of the local desert and its suggested sub-division into agro-ecological zones, we now come to the crux of the problem - introduction. What we have learned from the above survey is not only the urgent need for introduction, but also the location of the zone that is destined for it. Introduction is a very common term in the agricultural vocabulary. Almost all our cultivated plants are aliens, imported from near and far. Introductions in forestry, horticulture, vegetables and field crops are as old as agriculture.

Introduction into dry deserts from dry deserts is not very popular because it is not as simple as superficially thought. The question that will engage us for a time is the theoretical base of desert introduction. This will require departure from the main topic from time to time.

a. Homoclimatic Deserts

The five continents display large stretches of temperate or tropical deserts. Altogether, about 30 percent of the continents are desert in the wider sense. Quite a number of deserts are ecologically analogous or even homologous, especially in regard to moisture, and yet they are completely different from one another in their biota (though having many genera and families in common). The reasons for this biotic difference is mere chance and is directed by evolutionary history. Their biota may be well adapted to their environment but have not been conditioned by the latter in their taxonomic make-up. This is seen from the fact that despite their taxonomic differences, they display, in their different regions, similar morphological and physiological devices to fight drought. The question as to why different floras occur in homoclimatic regions is thus an historical one. Deserts and their floras are very old. The antiquity of deserts has been much discussed in the literature, especially by Soviet botanists such as Iljin[4], Lavrenko[5] and others. Evidence for the antiquity of the desert flora can also be found in the occurrence in the deserts of the Old and New World of a considerable number of genera and families, while lacking completely species shared by both of these regions. This indicates that the floras of these regions must have had a common stock very early in the geological history of the Gondwana Land before the continents moved away from one another to such an extent as to obstruct dispersal. Later, the flora of each continent had enough time to differentiate along its own lines. Examples of such genera are Ziziphus, Capparis, Prosopis and Acacia which are represented in the Neotropis and Paleotropis by different species.

Finally, the antiquity of the desert flora is shown by the occurrence of plants with highly elaborated morphological and ecological features and devices for survival which required geological ages for their evolution.

The biota of these regions have retained their specifity during geological ages without interchange with others in different continents and even in the same continent, despite the homoclimatic nature of the regions. The reasons for this isolation were obviously mechanical, namely, the inablity of nature to overbridge obstacles preventing long distance seed dispersal necessary for floristic interchange.

This long-lasting hinderance has now been removed by man who has become the most efficient seed dispersal agent. Man has shown the highest ability to carry out interchange of species between remote regions and so opened the way for large-scale introduction.

There is, however, a serious limitation to these introduction possibilities. Plants are ecological entities endowed with environmental requirements. No introduction can be made without possessing full knowledge of these requirements. The existence on the globe of homoclimatic regions opens, without doubt, enormous possibilities for the introduction and interchange of species of various homoclimatic regions.

So far very little has been done in the direction of interchanging desert plants of homoclimatic regions.

b. Introduction: Its Necessity and Its Reasons

Why introduction? It is now quite clear that the enormous phytomass which is so striking on the hills, mountains and plateaux of the so-called "green hammada" zone is totally unusable for man and herd. The bulk of this phytomass consists of scores of dwarf shrub species. The reasons for this depletion of plant cover are, first of all, the fact that the flora and vegetation of the Negev have been subjected to destructive action since the advent of man who has roamed in this surrounding for a million years and who, for understandable reasons, preferred open terrain for dwelling and hunting. No wonder the deserts of Israel; as those in the adjacent countries, have become extremely impoverished in their vegetation. It is obvious that these deserts were much richer in arboreal and otherwise useful plants than they are today. Evidence for this is vestiges of such elements, comparisons of fenced versus free vegetation areas, and, above all, facts and findings of more useful plants in homoclimatic deserts which have not been accessible to man or have only suffered relatively little interference.

Additional harm has been afflicted to these deserts by invasion and replacement of the destroyed, useful plant inventory (edible pastoral, industrial and arboreal) by highly aggressive and stubborn, inexpedient plants which waste the water resources of the deserts, such as species of Atriplex, Tamarix, Thymelaea and others. These kinds of "weedy" plants will never leave their positions unless removed by man.

All the above facts call for introduction and replacement trials of the immense uneconomic phytomass of our deserts by a more promising inventory. For this restoration two methods are available. The first is the amelioration of the vegetation by returning to it the lost inventory. There are, as a matter of fact, scores of grasses and leguminous plants which are still extant in the local deserts, though very rare. These plants must be collected and propagated for the purpose of <u>replacing</u> many of the dominant "useless" plants. The methods of this restoration work should be learned first under nursery conditions and then in the field. This restored vegetation requires much care, especially in balancing the equilibrium between the individual species. The second stage is the import of useful plants from other homoclimatic deserts which are still rich in useful plants since their vegetation has not been interfered with by man, at least not to such an extent as ours. This should be done on a scale proportionate to the problem posed of utilizing the huge zone of "green hammada" towards its conversion into pasture land. The world's deserts are now open to serious investigation in this direction and we are commended to make use of this in a planned, thoughtful and scientific manner.

There is now very efficient communication in the area of plant exchange. Our steps should be directed first to the Middle Eastern countries, to the world of grasses and legumes inhabiting the area between Sudan and Rajestan, to Iran - Middle and Central Asia. Floras and checklists are now available for all these countries. The deserts of Australia are among the richest in plants. The flora of Karoo-Namibia has much to offer our deserts. However, we are better acquainted with the southwestern deserts of North America, which are easily accessible and are no doubt one of the richest sources for introduction. The Atacama desert is less known to us as to its useful flora, but it too should be studied.

All in all, the problem of finding use for the tremendous extent of our "green hammada" zone is the cardinal point in revival or utilization of the desert and should be given first priority as a main subject for desert research. There are other remedies for this abused land in addition to pasture.

V. REFERENCES

1. Atlas of Israel (Jerusalem 1956).

2. M. Evenari and G. Orshansky, The Middle Eastern Hammadas, Lloydia 2, 1-13 (1948).

3. M. Evenari, L. Shanan and N. Tadmor, The Negev, The Challenge of a Desert (Harvard University Press 1971).

4. M.M. Iljin, Flora pustini tsentralnoy Azii, yeye proishozhdeniye etapi razvitiya. In: Materials on the History of the Flora and Vegetation of the URSS 3, 129-229 (URSS Acad. Sci. Press, Moscow-Leningrad 1958, in Russian).

5. E.M. Lavrenko, Osnov'ite chert'i botanicheskoy geography pustin' Evraziy i Severnoy Afriki (Komarovskiye Chteniya 15: 1-167. URSS Acad. Sci. Press, Moscow-Leningrad).

6. M. Zohary, Plant Life in Palestine. (The Ronald Press 1962).

7. M. Zohary, Geobotanical Foundation of the Middle East (Gustav Fischer Stuttgart - Swets et Zeitlinger, Amsterdam 1973).

8. M. Zohary and G. Orshan, Ecological Studies in the Vegetation of Near Eastern Deserts V. The Zygophylletum dumosi and its hydroecology in the Negev of Israel. Vegetatio, 5-6: 340-350 (1954).

CONTRIBUTION OF ANIMAL STUDIES TO HUMAN SETTLEMENT IN THE DESERT

DANIEL COHEN

I. THE ARENA: THE DESERT ECOSYSTEM

The deserts and the marginal lands surrounding them are not usually thought of as areas of animal production, but in fact, the world history of nomadism centres upon the arid zones and the movement of men and their animals across these vast tracts of land. This is true whether we think of the yaks and ponies of the Mongols, the sheep, goats and camels of the Arabs, the llamas and alpacas of the Indians of the Andes, or the beef cattle of the cowboys of the American west. Thus, human development in arid zones has always been intimately tied up with the development of animal production. This has frequently involved a long-range process of selection of the most promising species and traits for the propagation and survival of domesticated animals in the desert ecosystem.[1]

The delicate and often precarious life balance between man and animals in this demanding ecology has frequently led to drought or stimulated expansion and over-grazing and the subsequent starvation of livestock. This pattern has been cyclical in some parts of the world and progressively destructive in others (e.g., such as the Sahel). The growing demand of mankind throughout the world for food, due to the population explosion, has in turn placed greater demands upon livestock production everywhere, including the arid zones. The result has been the continuous conversion of marginal lands into true deserts through overgrazing, especially in Africa and the Middle East. This in turn has led to the search for alternative species which could be economically raised for food without the destruction of the ecosystem. Population pressures have also increased the pressure for further colonization of desert areas. As a result, man has unexpectedly intruded upon certain areas which may be the natural foci of zoonoses (diseases transmissible from animal to man) and thus a potential threat to human health. On the other hand, the study of wildlife in desert areas can be extremely useful as a source of information concerning the problem of adaptation to desert environment or by indicating environmental factors harmful to man and animal in the desert ecosystem. [2]

II. ANIMAL STUDIES CAN HELP

Animal studies or enterprises can make a positive contribution to development within arid zones without entailing permanent destructive alteration of the ecosystem.

The contribution of animal studies to the health of man in the desert can be subdivided into four areas of veterinary medical interest:

- Animals as sources of health hazards for man;
- Animals as monitors of environmental hazards;
- Animals as models for the study of desert diseases or physiological stress;
- Animals as providers of high quality food for peoples of the desert.

a. Animals as Sources of Health Hazards for Man

The study of animals as sources of human disease antedates veterinary medicine itself, and the recognition that animals play a role in the plagues of mankind is as old as the golden images of mice modeled by the Philistines to commemorate the epidemic of plague that hit Ashdod in 1320 BCE and resulted in 50,000 deaths in the Bet-Shemesh area alone. [3] Surveillance of zoonoses of the desert are not new either, and there have been very many fine studies concerned with the occurrence of such diseases as *coccidioidmycosis* in the lower Sonora region of the United States, *brucellosis* and *hydatidosis* amongst the desert nomads of the MIddle East, rabies in Iranian wolves and *leishmaniasis* in the Jordan Valley, to mention only a few. [4,5]

In a list of the zoonoses with which we ought to be concerned, I should include: *Rift Valley fever, anthrax, brucellosis, plague, relapsing fever, spotted fever, tuberculosis, ringworm, leshmaniasis, cysticercosis, hydatidosis, larvae migrans, toxoplasmosis, rabies, W. Nile fever, salmonellosis, leptospirosis,* and *Q fever.*

In studying the zoonoses, it is important not merely to record their presence, but also to identify their geographic distribution and natural foci and to undertake preventive measures to minimize their harmful effects. In this regard, the work of Soviet academician Pavlovsky, as recorded in his book "Natural Foci of Transmissible Deseases," is an excellent guide to the types of studies to be performed.[6]

This has already been done in the studies of *leishmaniasis* in Israel, in the studies there of malaria and the *arbo-viruses.* In this respect, the Negev offers some excellent opportunities for before-and-after studies of desert areas scheduled for settlement, insofar as wildlife reservoirs and insect vectors of zoonoses are concerned.

The Negev and northern Sinai also offer excellent opportunities to study the influence of the zoonoses on traditional nomads and on nomads now in the process of transition to more sedentary life styles. The study of the Bedouin in regard to specific zoonotic problems (such as several of those enumerated above) could serve as a model for evaluating their socio-economic importance to the Bedouin of the entire Middle East. The close association of these people with their flocks makes the zoonoses an area of vital concern. One of the mysteries worth pursuing is the story of *hydatidosis*, a tape-worm cyst that grows in the human liver to the size of a cantaloupe and is found in 10% of Bedouin sheep and cattle, and in almost every camel slaughtered.

One of the little anticipated consequences of man's penetration of the desert ecosystem is the subsequently increased human involvement with venonous animals, particularly snakes and scorpions. At present, this is a matter of some importance to the military and to those concerned with Bedouin health, but it ought to be anticipated as a predictable consequence of the establishment of new settlements in the desert, and appropriate public health measures taken.

b. Animals as Monitors of Environmental Hazards

A second major area of veterinary medical interest is the utilization of animals as monitors or sentinels for the detection of environmental factors hazardous to human health. The monitoring of animals for the detection of the presence of certain zoonoses is already a well-established practice and follows two distinct lines:

- monitoring of wildlife and domestic animals in their normal habitat; and
- stationing such animals in a particular environment to serve as sentinels.

This latter practice has been used extensively in arbo-virus surveillance and may be useful in monitoring some of the zoonoses mentioned previously in areas of the Negev scheduled for development. In a monograph ("Health Hazards of the Human Environment")[7] the World Health Organization indicated a number of examples illustrating how the incidence of spontaneously occurring diseases in animals has furnished pointers to possible environmental effects on human health. Among the examples offered were those of farm animals serving as indicators of environmental contamination following the radioactive fallout after an accident at Windscale in the United Kingdom; the use of pet animals as an indicator of harmful air pollutants in studies in Pennsylvania; the study of birds of prey which were seriously affected by DDE*, a breakdown product of DDT, etc.

The authors of the WHO monograph also discussed the relation between the discovery that bracken fern can produce bladder cancer in cattle and a search for a similar agent as a cause of stomach cancer in man in the United Kingdom. They remind us that *aflatoxin* was first called to our attention by an outbreak in turkeys before medical research started to look for its possible effects on man. The paper mentions the observations being made of various infectious and non-infectious agents on the production of congenital malformations in animals. I am reminded of the interest in *arthrogryposis* of sheep and cattle and its possible zoonotic implications for man. Lastly, the monograph reminds us that many of the first indications we see of the presence of a chronic chemical poisoning, such as *fluorosis* or *selenium* poisoning, are first noted in animals. It confirms that animal surveillance can be extremely useful in detecting carcinogens, teratogens and mutagens in the environment, because animals (unlike man) live within a restricted environmental area for most of their natural life span and, given their shorter life spans, they respond more quickly to pollutants and other chemicals than man does.

Animals will, therefore, be useful in monitoring the effects of industrial pollution in the new industries to be introduced into the desert. In this respect, one need only recall that the *Minamata* disease, caused by a mercury compound released into a bay in Japan by a plastics factory, first produced disease in the area's cats, years before its human helath implications were noted.[8] It would be a good idea to plan a system of environmental monitoring around

* Dichloro-diphenyl-ethylene

industrial plants and power stations, the system to include animal sentinels.

Such sentinels need not only be stationed in the field. Appropriate long-term exposure studies could be established in siumulated environmental conditions in laboratories. In this regard it must be pointed out that the world currently lacks sufficient animal testing facilities which could make such tests feasible.

The key to the successful utilization of animals as indicators, monitors or sentinels lies in the utilization of trained veterinary pathologists and clinicians.

A second problem area is that there are a variety of species including lagomorphs, canines and primates which are not available in adequate numbers for biological research. The world-wide utilization of primates for studies of carcinogenicity has rapidly depleted this wildlife resource, and the U.S. and Europe are now turning to primate breeding facilities to meet the need. I believe the desert offers an excellent area to develop primate breeding facilities to meet the world's research requirements. In the desert we have the climate and space to provide breeding facilities either in the open or with minimal shelters. The desert offers an excellent area for maintaining any animal raising facility which must be kept relatively remote and off the beaten track of human travel.

As we settle the desert we shall also expand the population of animals which are to be found in the region; not only the domesticated beasts, to be raised for food and transport, but also those pet animal species and commensal species of rodents and birds which will share our environment. They will share with man the effects of aridity, salinity, temperature variations and the particular effects of the chemical environment which our new life in the desert will demand. It will be useful to monitor these animals, particularly for the long-term effects of certain chemical components of our environment.

c. Animals as Models for the Study of Desert Diseases or Physiological Stress

The search for animal models of human disease should similarly be organized along more deliberate lines. Only 4 of the 243 genera and 3 of the 32 families of rodents and lagomorphs are used in biomedical research. There are 750 genera, including 7,000 species of small mammals, which have rarely or never been used as animal models.[9] We have known, for example, for many years that certain metabolic diseases associated with obesity were readily produced in certain wild mammals, particularly in species adapted to live in the deserts. Excess feeding of unlimited quanitities of high caloric food to such species or certain caged desert rodents (e.g., *Psammomys, Acomys* and *Ctenomys*) produced obesity, diabetes and impaired vision.

The adaptation of various wildlife species to laboratory conditions is problematic and depends on a careful study of the particular environmental conditions required by each species for breeding. For example, the construction of the cage must be such as to take into account the breeding habits of the animal concerned. Certain oppossums require a cage to allow the male to breed while hanging by its tail, and the cottontail rabbit needs space for the male to make a dash and leap before mating. We, therefore, require a group and facility whose primary function would be to develop and adapt various wildlife species to serve as animal resources for medical research. This group must span the gap currently existing between the observations of zoologists, on the one hand, and the needs of the medical community on the other. Veterinary medicine is uniquely trained to provide this middle ground.

In 1977, we established a veterinary hospital whose major function is to search for those spontaneously occurring diseases of animals which are peculiar to the desert environment, either quantitatively or qualitatively. The hospital receives clinical material from all over the country and monitors all of these animals for a variety of clinical and laboratory conditions and markers unique to desert environment. One project under way for example is the screening of urine and serum of animals for biochemical markers of metabolic defects which may be peculiar to desert conditions. This project is being done in collaboration with the University of Pennsylvania which has already examined animals from the Arctic and other ecological regions. The hospital's main functions therefore, are epidemiological and model building. Much effort is put into continuing education for veterinarians to enhance their diagnostic capacity and to increase our referred case load of interesting cases. In 1978 we opened a Camel Clinic which treats not only the Bedouins' camels but their other animals as well. A serum and data bank of all animals seen is maintained for future studies. This bank will be made available to researchers throughout the world who will be interested in biological and immunological research on animals of the desert.

Unfortunately, the establishment of the Camel Clinic comes at a time of declining interest in this animal in Israel from an agricultural point of view. We have watched with regret the decline in numbers over the past few years of camels in the Negev, although they are still an important animal in northern Africa and the Middle East. This animal is already adapted to the desert, does little destruction to the ecosystem, and is a multipurpose animal. The camel is an acceptable source of meat and milk, power and transport throughout Africa and Asia. Its development, limited by problems of parasitic and microbiologic infections and problems of reproduction and nutrition, is a manageable veterinary problem worthy of study. We intend to establish a centre to undertake over the next decade a systematic examination of all the anatomical, physiological, clinical, pathological, microbiological, parasitological and reproductive aspects of the camel with a view to enhancing its utilization as an animal model and as an economic resource. The Negev offers unique opportunities in space and ecology to study this animal in its entirety within its natural habitat and close to a scientific centre. This Camel Study Centre could have far-reaching socioeconomic implications for our entire region.

d. Animals as Providers of High Quality Food for Peoples of the Desert

All of the nomadic peoples of the desert have been, traditionally, primarily dependent upon their animals as the major source of their protein as well as of their clothing, shelter, transportation and defence; the close association between man and animals is most pronounced among the world's desert populations.

Most nomadism takes place in the marginal lands bordering on the true desert. It is dependent for its success on the mobility of its populations and on their ability to follow, without excessive overgrazing, the widely dispersed areas of sparse vegetation and water holes. Maintaining the ecosystem is essential to the nomads in the long run. However, at present, due to the pressures of over-population and world protein shortage, an expansion of the domestic herds and consequently, overgrazing have entailed marginal land erosion and a dramatic increase in man-made deserts. The same process which has been underway in the southern borders of the Sahara and which recently was shown to involve an extension of that desert of approximately one mile per annum, today threatens the Negev and northern Sinai where the expansion of sheep and goat herds by the Bedouin are producing rapid denudation of the local vegetation and a rapid conversion of half of the land mass of Israel into a true desert.

One of the solutions to the problem of maintaining the ecosystem while simultaneously expanding protein production is the development of appropriate alternative populations which can be harvested for food. Such proposals are under active study in Africa and should be studied in Israel as well.[10] Certain antelope species have the ability to withstand heat and water deprivation in excess of anything noted in domestic animals. Thus, animals such as the *addax* can go for years without water, and many animals such as the *Beisa Oryx* and *Grant's Gazelle* can conserve water by varying their body temperatures by as much as 10°C daily, without ill effects. Antelopes of the desert have a special anatomical mechanism which allows for the cooling of the blood before it reaches the brain. Antelope species have been shown to browse on species of bushes which cattle will not eat. These bushes are hygroscopic and at night, when the relative humidity is high, they increase their water content. Antelopes browse at night, cattle and sheep by day. These animals also conserve their body water more efficiently. The urine, faeces and milk of the antelopes are more concentrated and their sweat is minimal. In turn, the ability to store the heat provides the antelope with the ability to withstand the night cooling. Cattle and sheep, by their daily requirement for water and by their fastidious requirements of particular sweet grasses, are not as successful in utilizing the natural vegetation as are several varieties of the antelope species. In the Serengeti National Park (Tanzania) it was noted that four species of wildlife could browse off the same plant without interfering with each other, whereas cattle would crop the entire plant, whether or not they utilize a specific part. Antelope species have a better reproductive rate under desert conditions and mature earlier than cattle. Species such as the *Springbok* and the *Eland* have already been intensively studied as a source of animal protein which is superior to cattle in desert areas, while at the same time helping to conserve the environment. In a country which already makes maximal use of its water resources, there is much to be gained by turning to large-scale raising of animals (including game birds) which are less dependent on water and which can make better use of local vegetation. We must still search for the particular species which are best suited to the unique conditions in desert areas.[11]

Lastly, something should be said for the development of the indigenous camel as a more productive source of meat and milk. In most respects, the camel is a product of centuries of careful selection, primarily as a beast of burden, for transport. In this respect, the camel does not differ too much from the Zebu cattle of India, bred primarily to pull a plow and a wagon rather than as a source of meat. It has already been shown in many parts of the world, including India, that proper breeding and selection can increase the quantity of meat and milk of these same animals dramatically. The camel represents a similar resource - already present in our area, acceptable as a source of meat and milk, which could be dramatically improved through research. Over 20,000 metric tons of camel meat are consumed each year in Cairo alone and a large potential market is available for this animal throughout our region.[12] There are over 14-15 million camels in the world, about 11 million of which are dromedaries. About 6 million of them are to be found in 3 countries, Sudan, Somalia and Ethiopia.[13] But the major markets are to be found in the urban centres of the Arabian countries to their north, where they are considered a poor man's beef in a world of many poor men.

III. CONCLUSION

The desert can be made to make its own contribution to the world campaign against hunger and protein malnutrition with all its attendant medical problems and mental impairment, without in turn contributing to that disastrous destruction of the environment which is taking place today in many regions of the world. I have no doubt that the proposed research - including the suggested animal studies, breeding and testing centres - will make major contributions to the development of the Negev, with important implications for the development of the entire region as well as for the deserts of the world.

IV. ACKNOWLEDGEMENT

This article is a revised version of an earlier paper published in Kidma, Israel Journal of Development No. 8/ 1976 (Vol. 2, No. 4).

V. REFERENCES

1. Man, Culture and Animals, ed. A. Leeds and A.P. Vayda, American Assoc. for the Advancement of Science, Washington, D.C. (1965).

2. A. Riney, Science Journal, 5A(6):32 (1969).

3. K.F. Meyer, in Diseases Transmitted from Animals to Man, ed. T.G. Hull, Charles C. Thomas, Springfield, Ill., P. 527 (1963).

4. C.W. Schwabe and K. Abou Daoud, Am. J. Trop. Med., 10, 374 (1968).

5. CDC Proceedings of Symposium on Coccidioidomycossis held at Phoenix, Arizona, Feb. 11-13, 1957, U.S. Dept. of Health, Education and Welfare, Communicable Disease Center, Atlanta, Publication No. 575, USDHS (1957).

6. E.N. Pavlovsky, Natural Nidality of Transmissible Diseases, Univ. of Illinois Press, Urbana (1966).

7. Health Hazards of the Human Environment, World Health Organization, Geneva (1972).

8. L.K. Bustad, J.R. Gorham, G.A. Hegreberg and G.A. Padgett, JAVMA, 169(1):90 (1976).

9. J.R. Held and M. L. Kuns, Meaningful Scientific Use of Indigenous and Exotic Wildlife, II International Symposium on Health Aspects of the International Movement of Animals, Scientific Publication No. 235, PAHO, Washington, 19-25 (1972).

10. M.A. Crawford, Vet. Rec., p.305, March 16, 1968.

11. A. Horthoorn, World Rev. of Animal Prod., IV(19-20), 120 (1968).

12. H.E. Nawito. M.R. Shalash, R. Hoppe and A.M. Rakha, Reproduction in the Female Camel, Animal Research Inst., Cairo (1967).

13. Animal Health Yearbook, F.A.O., Italy (1976).

ALGAE PRODUCTION FOR BIOMASS IN ARID ZONES

AMOS RICHMOND

I. INTRODUCTION

The primary source of all man's food and organic raw materials is solar energy. As humanity expands and its demands persistently increase, man's existence may, in the final analysis, depend on how efficiently he learns to use solar energy to produce the required quantities of food, energy and organic materials.

For several reasons agriculture is very inefficient in this respect and most conventional crops utilize about one per cent of the sunlight that falls on them.

Algae on the other hand, promise important advantages in improving the efficiency of solar energy utilization for valuable biological products. These are aquatic plants, which range in size from large multi-cellular oceanic species of over 30 m in length, down to microscopic unicellular forms, comprising several thousand species which may be grown under a wide range of conditions.

Algae production represents an extreme approach in modifying desert conditions for the growth of plants. In this work, I shall relate to the issues involved in the production of algal biomass by elaborating on three major questions:

1. What are the distinct advantages of algae over land plants, particularly in arid zones?
2. To what uses can algae be put?
3. What are the basic biological and technical aspects that should be elucidated to facilitate commercial production?

II. BIOLOGICAL AND ENVIRONMENTAL ASPECTS

The general concept of algae production in arid zones is plain: it involves the use of the high rate of solar irradiance and the high temperature prevalent in most deserts throughout the year, the wide open spaces and saline water (brackish or even sea water), i.e., water not suitable for the production of most useful land plants, to grow algal plant-biomass. Thus, resources which are prevalent in many arid zones, and which in effect post limitations to growth and development of conventional agriculture, are conducive to the growth of many algae species, bestowing a specific advantage on alga-culture in many hot and arid areas.

In addition to their tolerance of saline water, many species of algae have other distinct advantages over conventional agricultural plants. Algae can be grown in a continuous culture, and thus it is practically possible to grow them in such a way that the concentration of biomass will always be in the optimal range for the absorbance and use of solar irradiance. Such a feat cannot usually be accomplished with land crops in which there is a long delay following harvesting before new vegetation covers the ground optimally. Another important advantage in alga culture is that unlike conventional crops, the entire plant body of the alga is harvested and used. This is because algae are water plants which are under no pressure to allocate special resources for growth of special organs, in contrast with terrestrial plants which require accessories such as a root system for anchorage and a branching system to carry foliage.

Still another advantage for alga culture is that the nutrient requirements may be easily maintained at optimal level. Clearly, three classes of plant nutrients are distinguishable; mineral elements, carbon dioxide (CO_2) and water. If one of these factors is not optimally available, the growth of the plant is limited. Thus, production of plant biomass depends on defining and alleviating the limiting factors for growth and development. In algae it is readily possible to maintain growth conditions in which there is no nutrient limitation. By comparison, the growth of well-watered and well-nourished field grown agricultural plants, exposed to high irradiance is soon limited by shortage of CO_2. In contrast, bicarbonate and CO_2 can easily be introduced into the algal medium to alleviate carbon limitation to growth. Algae such as *Spirulina* may serve as an example. They can be grown at pH 10, at which on the one hand pH absorption of CO_2 from the air into the medium takes place and on the other, CO_2 which is supplied to the medium is completely absorbed in the bicarbonate-carbonate system, without any loss.

In an intensive, correctly maintained algal system, therefore, the only limitations to growth may be environmental, i.e., temperature and light. Regarding temperature as a limiting factor, this could be to a large extent eliminated by the proper selection of the production site. Indeed, another important point regarding the suitability of algae grown in the desert is that many useful strains of algae require high temperature, with an optimum range of 35-38°C. Thus, the prevalent high temperatures in arid zones which are growth-limiting for most conventional crops represent the optimal for suitable algae.

When nutrients and temperature are not limiting, growth and production become limited by light. Consequently, locations which are rich in solar irradiance throughout the year, such as most of the desert areas of the world, have a distinct advantage. Clearly, this combination of conditions permits the plant cells to operate close to their intrinsic maximal photosynthetic efficiency.

Today most scientists believe that the theoretical maximum energy efficiency for the photosynthetic reduction of CO_2 to carbohydrates is ca. 28%[1] even though recently Pirt *et al.*,[2] brought evidence that it could be as high as 50%. The photosynthetic apparatus of all plants is such that they can utilize only light in the wavelength range of 400 to 700 nm. This irradiance is known as "photosynthetically active radiation" (PAR) and amounts to about 48% of the total solar irradiance which reaches the earth's surface. Thus, for the efficiency of total solar energy conversion, it is necessary to correct photosynthetic efficiency by PAR. Accordingly, if the maximal photosynthetic efficiency is 28%, then the maximal efficiency for the utilization of solar irradiance will be ca. 12%. This figure may be taken as the maximum for algae, allowing for the fact that an algal culture maintained at optimal population density absorbs practically all the PAR reaching its surface and that the extent of light reflection and photo respiration in these C_3 plants is minimal. There must be some loss due to respiration, but this is found to be relatively small, i.e., ca. 10% of the photosynthetic activity in cultures with correct cell density. During the night, when the temperature in the pond declines and energy is needed only for maintenance, respiratory losses are slight. Therefore, I propose that it is correct to set ca. 10% photosynthetic efficiency as an attainable goal in the production of algal biomass. Of course, if the data reported by Pirt *et al.*, are generally substantiated, then perhaps efficiency as high as 18% in total solar energy utilization could become the goal of biomass producers. For comparison, biomass yields which correspond to energy conversion efficiency of up to 5% on a year-round basis are considered maximal for higher plants grown under field conditions,[1] one percent efficiency being considered an average.

Algae then, may be among the most efficient plants in solar energy utilization and may produce the highest output rate in dry weight and protein per unit area, when the sole limiting factor for growth is light. Thus, the most important single issue in the practical understanding of the biology of biomass production is the study of light as a limiting factor in outdoor cultures.

Due to self-shading,[3] the meaningful parameter in studying the effect of illumination on the culture is the integrated radiant flux incident on each algal cell. Outdoors, this parameter is affected by three factors: (1) light intensity; (2) the population density in the pond; and (3) the dark/light cycle to which an average single cell in the culture is exposed. The latter is affected by the extent of turbulence, the depth of the medium and the population denisty, as will be further elucidated.

The inter-relationship between light irradiance and population density is illustrated in Table 1 which depicts a typical pattern of distribution of incident light throughout the depth of a pond of *Spirulina platensis*. Clearly, the distribution of light is a function of cell density.

TABLE 1, Percent distribution of incident light throughout the depth of a *Spirulina platensis* pond.[4]

Pond depth (cm)	Cell concentration (in optical density units; 560 nm)		
	0.10	0.22	0.40
1	57	47	35
2	49	33	16
3	43	18	3
4	38	8	0
5	29	2	0
6	18	0	0
7	2	0	0
8-15	0	0	0

100% incident light = 2,300 microeinsteins m^{-2} sec^{-1} (400-700 nm)

Data in Table 1 indicate that when cell density was relatively high (0.40 optical density at 560 nm or ca. 800 mg dry weight per each litre in the culture), only the upper 3 cm of the pond, or about 20% of the cell population, received some light. Accordingly, about 80% of the cells were practically in complete darkness at any given moment. Even when cell concentration was halved, solar light did not penetrate beyond a depth of 5 cm and over 60% of the cells were left in complete darkness at any given time. Furthermore, in cultures of very low cell densities, such as exhibit the maximal specific growth rate but could not be economically maintained in the pond, light penetrated to less than half the pond depth, and more than half the cell population was exposed to complete darkness at any given instant. Thus in outdoor ponds, the extent of mutual shading which is a function of population density, cell size and pond depth - is the major factor determining the amount of solar light available to the photoautotrophic cells in the culture. Another point is that due to the turbulent flow which is maintained in the pond, each cell in the culture is exposed to a light/dark cycle which may take a few seconds to many minutes to complete as each cell travels back and forth from the upper, illuminated layer of the pond, down to the lower and much larger, unilluminated layer. The light/dark regime to which each cell in the culture is thus exposed has hardly been investigated, but I believe has an important effect on the growth rate and photosynthetic efficiency. The nature of the light/dark regime depends on the intensity and duration of solar irradiance, the depth of the pond, the population density and the extent of turbulence in the pond. The latter effect has been studied by us in some detail and is referred to below.

Since the net output of biomass is a product of both cell density and the specific growth rate, and since these parameters are negatively and essentially linearly related, it is clear when the system is only light limited that maximal output may be achieved at some optimal cell density.[5] This is shown in Figure 1.

Also evident from this figure, the output rate is decisively affected by the extent of turbulence in the pond. One interpretation[5] of this marked effect is that it sets a more favourable light/dark cycle for each single cell in the culture, resulting in improved photosynthetic efficiency. It is evident that the greater the turbulence, the shorter becomes the duration of one complete light/dark cycle.

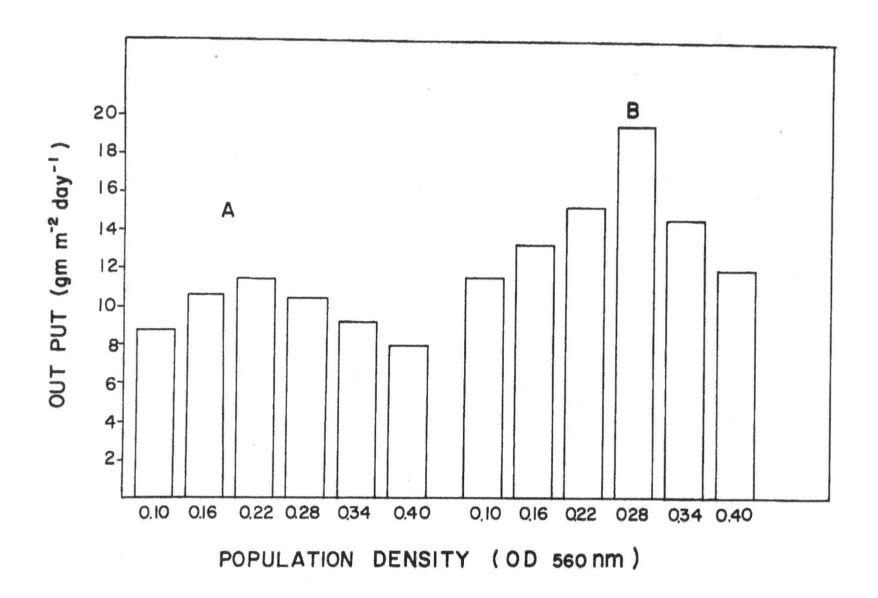

FIGURE 1, Output rate as effected by cell density and turbulence.

A = paddle speed 15 RPM
B = paddle speed 30 RPM

Also when irradiation is very high, e.g. 2,500 microeinsteins·m^{-2} sec^{-1}, cells placed in the very upper layer of the pond may suffer from over exposure to irradiance. Intense stirring would decrease the duration in the over-exposed upper layer. Clearly, increased turbulence affects increased growth rate which is shown in Figure 2. Support for our thesis that equates greater turbulence with improved light regimen for the photoautotrophic cell is that when stirring is enhanced, peak output of biomass is shifted, being obtained at a higher cell density (Figure 1).[5]

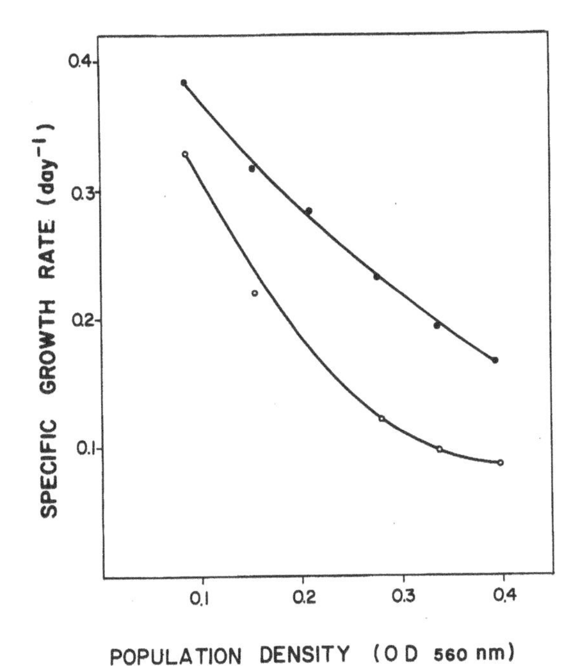

FIGURE 2, The influence of turbulence on the specific growth rate at various cell densities

●——● paddle speed 30 RPM (upper curve)
o——o paddle speed 15 RPM (lower curve)

The complexity of optimizing outdoor algal biomass production throughout the year is reflected in Figure 3, in which the effect of X (population density) on μ (the average specific daily growth rate) is delineated. Since μ is substantially modified by temperature, this relationship varied greatly throughout the year. Clearly, the more severe the temperature limitation on μ, the smaller its dependence on X, becoming hardly visible in mid-winter when the specific growth rate is very small (open circles, Figure 3). In the summer (open triangles, Figure 3), light became the main limiting factor for the output rate and close relationship existed between the specific growth rate (μ) of the culture and its cell concentration (X)[5].

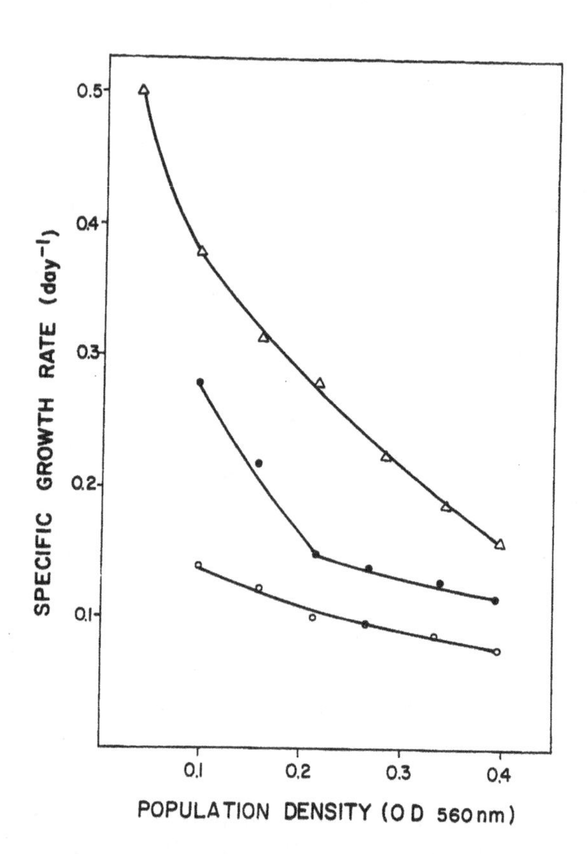

FIGURE 3, The effect of cell density on the specific growth rate[5]
Δ-Δ-Δ in August-September (top)
•-•-• in May-June (middle)
o-o-o in December-January (bottom)

In addition to the effect of light and temperature, a major biological question which concerns the mass production of algae is to what extent it is possible to maintain a unialgal continuous culture outdoors. In nature, there are some examples of algal species such as *Spirulina* which dominate in a body of water as a major photosynthetic species. In small-scale experiments, we found that a continuous culture of *Spirulina* was readily maintained throughout the summer. Cell density was kept constant by continuously filtering the excess biomass, while the effluent was returned to the pond and the volume of the medium was kept constant by daily addition of tap water. CO_2 was added to maintain the pH within a range of 9.5 to 9.8 and the nutrient level was maintained by analyzing three times weekly for PO_4 and NO_3, adding complementary amounts of the entire mineral make-up when supplementing nitrogen to the medium. Results clearly indicate that as long as temperatures in the pond did not decline below 20°C, *Spirulina* culture could be kept essentially clean of other biological species. The number of bacterial cells did not increase above 1×10^3 $m\ell^{-1}$. In addition, analysis of the daily specific growth rate did not reveal any signs of self-limitation in the pond throughout its continuous operation from April to October.

A crucial practical aspect of pond maintenance is to have means by which the relative performance or "well-being" of the culture in the pond can be constantly and readily evaluated. We, as well as others[6] found that the partial oxygen pressure in the pond during sunlight serves as a useful tool. Table 2 illustrates the effects of the intensity of solar irradiance and temperature on pond oxygen. All figures are based on data recorded hourly throughout the year. Most figures shown represent averages of many scores of observations, 10% of which were deleted from both extremes. All readings were taken at 1 p.m. when pond oxygen was usually on the increase.

TABLE 2, Effects of the intensity of solar irradiance and temperature on pond oxygen in *Spirulina platensis*[4]

	Pond Oxygen: % of Saturation					
	Incident Light - K lux					
Temperature °C	0-5	5-20	20-40	40-60	60-80	80-100
6-12	71					
12-18	87	95	109	115	145	
18-24	88	101	113	122	140	157
24-30	94	108	120	131	142	181
30-36	107	---	135	125	155	208

Clearly the highest pond oxygen was recorded at the highest temperature and irradiation, thereby illustrating that production of this warm water alga depended on both these environmental parameters.

III. TECHNICAL ASPECTS

Three major technical aspects are involved in developing commercial systems for the mass cultivation of algae. The first relates to pond construction, its shape, depth and system of mixing the algae-laden water, and most important, the type of lining. The second concerns separating the algal mass from the medium, and the third relates to the dehydration of the algal mass to facilitate product distribution and storage, A satisfactory solution to these three problems is essential before any large-scale cultivation can be economically pursued.

a. Pond Construction

Mass production of algae requires an inexpensive method of pond construction. Since the medium must be well stirred, shallow race channels about 20 cm deep are required. One possibility[7] is for the channels to be dug with a constant gentle slope in the direction of flow, assuring a rate of approximately 30 cm min^{-1} which should create sufficient turbulence to mix the algae adequately, an essential for the distribution of solar radiation to all the cells in the culture. At the end point of this channel maze, the algal medium would be lifted some 1.0 or 2.0 m, to be returned to the point of origin. This system, however, may not be suitable for algal species which may become damaged going through the pump.

A simpler and seemingly more feasible system of pond construction is being used by the Proteus Corporation of California which produces *Spirulina* commercially. There, the raceways are 200 meters long and 10 meters wide, making a 2000 m^2 pond unit. The raceways are lined with 2mm thick lining which is guaranteed to last 10 years. A set of paddle wheels across one channel provides sufficient stirring, i.e., producing a flow of ca. 30 cm min^{-1}. If the lining lasts for 10 years as expected, the capital outlay for such a pond seems economically feasible.

b. Separation of the Algal Mass from the Medium

Harvesting the algal biomass requires separation of the cells from the aqueous medium. Various methods are possible for this and the size of the algae to be separated governs the choice. In general, harvesting may be achieved by applying centrifugal force to separate the algae from the medium. This system however, is not economical and cheaper methods must be sought.

Separation may be greatly facilitated by introducing chemicals into the medium which induce the single cells to flocculate. A commonly used flocculant is aluminium sulphate which promotes separation by flotation and thus the foamed biomass can be easily filtered. However, the algal powder which is obtained after dehydration contains about 20% aluminium and therefore this form of separation may impose a difficulty for use as feed, unless the algae are further treated. A system designed to extract the aluminium from the dry matter by acidification and washing is feasible, but increases the cost of harvesting.

Obviously, if it were available at low cost, an edible flocculant would be very useful in the production of algal biomass for food or feed. Causing the algae to sediment, either by settling on suitable membranes or by the addition of proper agents is a further technique of separation. The sedimented algal mass may then be collected by various means. Today filtration appears to be the obvious choice for the first stage of separation from the aqueous medium. The type of filter and the process depend on the algae to be separated and on the production rates. Small micro-algae, such as *Scenedesmus* or *Chlorella* with a diameter of a few microns, pose great difficulty due to their rapid clogging of the filter. However, there is a report about a micro-algae harvester[8] based on a pre-coated paper belt running over a filtration drum. The process is based on a double fabric belt that sandwiches the algae-paper mat on both sides, so as to retain the paper fibres during one application of vacuum and water and then to separate the algae from the belt. The precoat and residual algae are then removed from the belts by water showers, the dispersed fibres are washed to remove residual algae and the washed fibre recycled to form a new pre-coat. If instead of a paper layer, an edible filtering material could be devised, i.e. made of some plant residue, then the produce of separation would be an edible mixture for animal feed.

Filtration of micro-algae growing in colonies that measure some 25 microns in diameter, such as *Coelestrum proboscileum*, is already possible. Using a machine developed for the food industry for thickening starch suspension, it was possible to concentrate a suspension of *Coelestrum* some 100-fold,[9] from 0.6 to 60 g ℓ^{-1}. However, at present, only the filamentus *Spirulina* is readily filtered on a 400 mesh commercial filter. Recent developments in the commercial production of Microstrainers which can filter particles as small as one micron, may provide the solution for harvesting single cell algae.

c. Dehydration of the Algal Mass

Dehydration forms a problem of major economic importance. Systems differ both in the extent of capital investment and energy requirements and have a marked effect on the biological value and taste of the produce, particularly with regard to green algae which have a cellulose encasement. One possibility is to dehydrate algal mass with a fine layer drum dryer, which yields an excellent product. In this method, an algal suspension of 6-8%, obtained by centrifugation, is ejected onto a rotating, steam heated drum: the algal cells are thus rapidly heated for a few seconds up to 120°C and the ensuing dehydration causes the cellulose encasement of the green algae to open up, thereby greatly increasing edibility and biological value of the powdered product[10] A double drum dryer between which drums a homogenized 5-10% *Spirulina* solution is placed is being satisfactorily operated by the Proteus Corporation in California.

Soeder and Mohn[9] proposed that dehydration may be successfully achieved by mixing algae with dry additives such as straw, sugar beet pulp, meal powder or grains, and thereafter pressing the mixture by extrusion to produce pellets instead of powder. In addition, they showed that dehydration expenditure is appreciably cut when the biomass is dehydrated to contain 10% rather than 5% water. Indeed, it is conceivable that certain mixtures could be kept successfully in a more hydrated form than is possible with the pure algal powder. Direct dehydration of algal biomass in the sun is feasible, but it is not recommended as it is harmful to quality and requires large drying areas. Nevertheless, a possibility so far unexplored is the harnessing of solar energy to create a source of heat for dehydration of the filtered biomass.

IV. USES FOR ALGA

The most obvious use for alga is as animal feed and as a human food supplement. A recent United Nations survey on the world food demand and supply has shown that about a quarter of the world's population (one billion people) are today getting less than the minimum calorie intake necessary for health.[11] It would take about 100 million tons of wheat equivalent, distributed throughout the world, to raise the diet from the starvation range. Arable land resources are quickly stretched to the limit. Thus, in the long term, significant additional food output could only result from increased crops per unit area. However, in the last 30 years, the increase in conventional crop yield per unit area has averaged less than 1% $year^{-1}$, while the world population has been increasing by over 2% $year^{-1}$. As a result, the chances today seem that by the turn of the century, the amount of food available per head will be significantly reduced.

The food potential of certain microscopic algae has been fairly intensively studied in the past few years. The blue-green alga *Spirulina* belonging to the family *Oscillatoriaceae*, is particularly interesting. Rediscovered by the academic world as recently as 1940, *Spirulina platensis* has been collected from the salty lakes and ponds along the northern shores of lake Tchad since time immemorial, sun-dried and eaten by the Kanembou people, now numbering 80,000[12]. *Spirulina geitleri* was collected and prepared similarly by the Aztec Indians at the time of Cortez' arrival in Mexico. *Spirulina* is an easily harvested multicellular filamentous alga of high digestibility and mild flavour which has been found to contain up to 70% protein of good nutritional quality.

Likewise, the quality of protein from the green alga *Scenedesmus obliquus* has been thoroughly investigated.[13] Considering any international nutrition-parameter, *Scenedesmus* and *Spirulina* compare very favourably with most common animal feeds, such as soybean extract and fish meal, as shown in Table 3.

The high content of proteins found in microalgae makes this product a concentrate by itself, having a significant advantage over conventional vegetable sources of protein which usually have a much lower protein content.

To improve the amino acid balance of algal protein, additional sulphur-bearing amino acids could be added. This is easily done by including the gluten of wheat or barley in the diet and in fact both the Kanembou and Aztecs used *Spirulina* with whole grain cereals.

TABLE 3, Nutritional values of standard feeds and algae[10]

Feed	% Protein	PER[a]	NPU[b]	BV[c]
Spirulina platensis	56	1.80	62	75
Scenedesmus obliquus	54	1.85	61	75
Soybean extract	50	1.82	60	70
Fish meal	57	1.82	66	79
Casein+methionine	90	2.50	86	92

a Protein efficiency ratio
b Net protein utilization
c Biological value

The nutritive value of this algae, as well as many other strains of algae, is amplified in that it has a relatively low percentage of nucleic acids (4%), as compared with high content of nucleic acid in bacterial protein. The mucoproteic membranes that separate the cells are easy to digest, unlike the cellulose cell wall found in many other nutritional algae; it is completely non-toxic, its lipids made up of unsaturated fatty acids that do not form cholesterol, perhaps making the *Spirulina* an interesting food item for patients with coronary illness or obesity.[14] The possibility of using algae as a human staple has been studied in the past decade. A significant insight was provided by the work of Hernandez, Gross and Gross,[15] who have investigated the effect of *Scenedesmus acutus* powder as a food additive in Peru. The authors believed that the high protein content, as well as the iron, vitamin B complex and carotene, along with considerations of previous positive tests on humans, made this powder an interesting product for combating protein-energy malnutrition and vitamin deficiencies. In a 4-week test, healthy individuals received a daily supplement to their normal diet, consisting of 10 g alga powder for adults and 5 g for children. Examination of blood, urine and faeces in the beginning and at the end of the study and also allergy tests revealed a satisfactory acceptance of the microalga. Likewise, slightly malnourished children around 4 years of age demonstrated a satisfactory food acceptance of the micro algae. More important, heavily malnourished and hospitalized babies who received ca. 500 mg. algal powder per kg. body weight showed marked improvement in their condition as demonstrated from psychomotoric and anthropometric data and from their blood chemistry profile. The weight gain was highly significant as compared to the performance of the same children tested before the algae were added to their diet. The authors concluded that their tests demonstrated the lack of toxicity from the consumption of algae as well as the good acceptance of *Scenedesmus* from the physiological point of view. The possibility of introducing *Scenedesmus* powder into Peru as a complementary food stuff was also investigated.[15] Five recipes for which there was good acceptance and seemed suitable for industrial production, were selected from various alga foods. In a mass test, 1745 Peruvians of middle and lower class who were tested for response to these five recipes yielded acceptability in the range of ca. 75-81%. In another study, Bhumiratana and Payer[16] reported on acceptability tests conducted with public school children in Bangkok. They pointed out how much the outcome of acceptability tests depend on the method used: green noodles containing 3.5 gm algae per meal were hardly accepted in the beginning of the test and repeating the test after another two weeks yielded only slightly better results. After two months, however, the "green noodles" were accepted by 98% compared with well known yellow noodles which did not contain algae. Bhumiratana and Payer also reported on clinical tests conducted in Thailand. Intake of 12-14 g algae per person did not influence the normal uric acid level in the blood plasma of Thai people. They indicate that such a daily amount would contribute a considerable quantity of protein to the present low intake in Thailand by persons suffering from protein malnutrition.

A significant nutritional attribute of *Spirulina* is its relatively high level of linolenic acid, indispensible for fish feeding which seems to give this algae a great advantage for pisciculture. Similarly, preliminary conclusions of growth experiments with Carp and Tilapia revealed that algae were very well accepted by these fish. Significantly, microalgae contain a high quantity of carotenoids, important for intense colouration of shrimps and certain species of fish.

Protein is only one of the several produces which can be commercially derived from algae. There are various chemicals which are already extracted from algae or which may become commercial products.

Today, three major algae products are extracted from marine algae[17]. These include alginic acid derivatives, carrageenin and agar:

Alginic acid is extracted primarily from laminaria and microcystis. The alginates are used for various purposes such as in the food industry, cosmetics and the textile and rubber industries.

Carrageenin, like the alginates, is a cell wall polysaccharide complex which gels in the presence of potassium and is used like alginates to stabilize emulsions and suspend solids in foods, as well as in the textile, pharmaceutical, leather and brewing industries.

Agar is a name used for a dried or gel-like non-nitrogenous extract from rhodophycean algae. It is used as a medium in the culture of bacteria, fungi and algae and also in numerous products in the food and chemical industry.

To a much lesser extent, marine algae are used for pharmaceutical purposes. Yet there is a growing interest today in finding a biologically active compound derived from marine organisms. A drug of some importance as an antihelmitic has been described[17] and a produce from the sea-weed laminaria was reported to have anti-hypertension activity.[18] Finally, there have been numerous reports of some antibacterial products from various algae.[18]

Recently much thought has been given to the possibility of using algae for the production of biofuels. Hydrogen and methane particularly have been discussed. It seems, however, that fuels are not yet sufficiently expensive to warrant their production from algae. In addition, with the present know how of biomass production, the energy used to produce and harvest the biomass may be greater than the energy obtained. Nevertheless, new ideas seem to be constantly brought up. One is by Pirt[19] who suggests using algae in a closed, air free system, through which passes industrial waste-CO_2. In such a system, the output of gas is some 80% O_2 and ca. two parts O_2 would be produced for each part of algal biomass. He suggests that the algal biomass would be fermented to methane and the residue recycled. An annual output yield of 80 tons dry wt. per hectare would yield some 40 tons methane and 160 tons of O_2.

A promising possibility seems to lie in products such as glycerol[20] and other alcohols, acids and starches[21] and preliminary experiments are being conducted to test the economic feasibility of deriving these products from algae.

Another outlet for algae is in biochemicals and various other natural products which today are required in much smaller quantities and could serve as by-products. These include pigments, vitamins, special lipids, steroids and other such natural products.

Finally, the human health market may become an important outlet for algae. In Japan, a powder made from the alga *Chlorella* is regarded as having remarkable health properties. Indeed, the very high protein content of an alga such as *Spirulina* - up to 70% - as well as its outstandingly high content of vitamin B_{12} [22] would make certain algae a promising food item for humans in general and vegetarians specifically. It has also been claimed that algal powder possesses various therapeutic qualities for healing gastric ulcers, various kinds of wounds and liver necrosis, in the regulation of blood pressure and in the prevention of decrease in leucocytes.[23]

In summary, consistent efforts towards optimization of algal biomass production, harvesting and product processing, as well as genetic improvements and basic research will yield in time, biotechnologies of significant economic importance. This would seem particularly useful in arid lands, in which populations are malnourished and in need of organic raw materials, but in which cultivation of plants by conventional methods is severely handicapped. In these areas algaculture in brackish or sea water has distinct advantages.

V. REFERENCES

1. J.A. Bassham. Increasing crop production through more controlled photosynthesis. Science 197, 630-638 (1977).

2. S.J. Pirt, Y. Kun Lee, A. Richmond and M. Watts Pirt. The photosynthetic efficiency of *Chlorella* biomass growth with reference to solar energy utilization. J. Chem. Tech. Biotechnol. 30 (1980).

3. H. Tamiya. Mass culture of algae. Ann. Rev. Plant Physiol. 8, 309 (1957).

4. Proc. of the Inter. Congress on Microalgae, Acre, Israel. The National Council for Research and Development, September (1978).

5. A. Richmond and A. Vonshak. *Spirulina* culture in Israel. Arch. Hydrobiol. Beih. Engeben Limnol, 11, 274-280 (1978).

6. E. Von Strengel and J. Reckermann. Methodische Vor arbeiten zur Messung der photosynthetischen Sauerstoff produktion in offenen Algengroskulturen. Arch. Hydrobiol. 82, 263-294 (1978).

7. A. Melamed. "Tushia" Engineering Co., Tel Aviv, (Personal communication).

8. Caldwell Cornel Engineers, Melbourne. Field testing an algae harvesting process. Report submitted to Bureau of Environmental Studies. Dept. of Environment, Australian Government, Canberra (1975).

9. J. Soeder and H. Mohn. Technologische Aspecte der Mikroalgen Kuller. Symposium Midrobieller Proteinge winnung (1975).

10. W. Pabst. Die Massenkultur von Mikroalgen. Kraft futter 58, 2 (1975).

11. J.D. Gavan, D.E. Hathaway. PAG Bulletin (United Nations) 7 (1-2), 5-27 (1977).

12. P.T. Furst. *Spirulina* Human Nature, 62-65 (March 1978).

13. W. Pabst. 1. Symp. Mikrobielle Proteingewinnung. S. 173. Weinheim Verlag Chemie (1975).

14. Durant Chastell and Sanches, Sosa Texcoco S.A., Mexico, FAO - Technical Conference on Aquaculture, Kyoto, Japan (1976). The Te *(Spirulina)* and the Aquaculture.

15. V. Hernandez, U. Gross and R. Gross. Some remarks about a testing programme for single cell protein (SCP) as food additives in Peru, on the example of the microalga *Scenedesmus acutus*. In a publication of Instituto de Nutricion, Tr. Tizon Y. Bueno 276, Lima 11, Peru.

16. A. Bhumiratana and H.D. Payer. 2nd Report on the production and the utilization of microalgae as a protein source in Thailand. (1972-73). Institute of food research and product development, Kasetsart Univ. Bangkok, Thailand.

17. A. Pletcher. F. Hoffman-La Roche and Co., Ltd., Basel Switzerland, in FEPA Kongress at Venedig, (1975).

18. A. Mutsui. The use of photosynthetic marine organisms in food and feed production. The International Conference on Bio-saline Research, NSF, Washington, D.C., (1977).

19. S.J. Pirt. Microbiol photosynthesis as a route to renewable sources of energy, food and carbon chemicals. Paper presented at IUPAC Conference, Fermentation Commission, Sept. (1979).

20. A. Ben-Amotz, Photosynthetic and Osmoregulation Mechanisms in in the Halophilic Alga *Dunaliella parva*. Ph.D. Thesis, Weizmann Institute of Science, Rehovot, Israel, (1973).

21. M.W. Pirt and S.J. Pirt, Photosynthetic production of biomass and starch by *Chlorella* in a chemostat culture. J. Appl. Chem. Biotechnol. 27, 643-650, (1977).

22. W. Shurleff, Sources of vegetarian vitamin B_{12}. Vegetarian Times, May/June, 31, 36 (1979).

23. Lab. Report No. 6020428 issued by Japan Food Research Laboratories, (1976).

HIGH YIELDS AND LOW WATER REQUIREMENTS IN CLOSED SYSTEM AGRICULTURE IN ARID REGIONS: POTENTIALS AND PROBLEMS

J. GALE

I. INTRODUCTION

This chapter is not a presentation of research results but rather a discussion of the possibility of developing an option for intensive agriculture in hot, arid regions.

Arid deserts are characterised by lack of fresh water, poor soil and high intensities of and high total radiation (about 8000 M joules m^{-2} yr^{-1} in Israel vs. about 4000 in central western Europe). The little available water is frequently brackish (Issar[1]). Even the most optimistic estimates put desalinised water at a price above that which would be acceptable for open field, conventional agriculture. The challenge, therefore, is to develop a system which uses very little water and with which a modern farmer can maintain a reasonable standard of living, using the available resources of the region, especially solar radiation. The high levels of solar radiation may be utilised not only directly for plant growth, but also indirectly for replacing fossil fuel used for heating greenhouses in more northerly climates.

One of the ways in which this may be accomplished is with closed system agriculture. By this we mean, essentially, a greenhouse type structure in which surplus solar energy is stored during the day and released at night. There are a considerable number of advantages of such a system: a) the high humidity atmosphere would reduce transpiration (plant water use) by a factor of between 5 and 10 and the transpired water could be re-cycled; b) extremes of temperature during the day and night would be prevented and use of conventional energy for heating and cooling would be much reduced; c) carbon dioxide supplementation is made feasible and d) optimal growing conditions could be maintained throughout most of the year. There are also many other advantages of growing plants in closed systems which are common to all the greenhouse industry, but are particularly pertinent under desert conditions. These include: prevention of wind damage and sand abrasion, ease of control of pests and, to a lesser degree, of diseases, and the possibility of closely controlling soil water and nutritional requirements.

The concept of closed system agriculture is not new. Israel patent No. 5528 was granted to E. Rappaport in 1952 and the concept has recently been proposed for protein production in arid regions by Bassham.[2] Elements of these ideas have been discussed at a number of recent meetings particularly in relation to energy saving in greenhouses (e.g., Solar Energy Conf. Int. Symp. Controlled Environment Agriculture).[3,4]

II. PLANT GROWTH CONSIDERATIONS

a. Water Balance

Before describing our approach to the engineering problem, it is appropriate to consider how a closed system can affect transpiration, whether the reduction of transpiration will be detrimental to plant growth and how CO_2 supplementation can affect plant development and yield.

The rate of transpiration, Q, is governed by a number of factors and can be described by the diffusion equation:

$$Q = \frac{e_1 - e_a}{r_a + r_s} \quad (1)$$

where e_1 is the vapour pressure of the leaf (determined mainly by leaf temperature), e_a is the vapour pressure of the surrounding air and r_a and r_s are the vapour diffusion resistances of the leaf boundary layer and of the stomatal apertures, respectively. It follows from (1) that if e_a is kept high as in the closed system, Q will be low. High levels of CO_2 partially close stomates, causing an increase of r_s (Meidner and Mansfield)[5] and hence, also tend to reduce Q.

The energy budget of the leaf (Raschke)[6] may be described by an equation of the form

$$Q_{abs} = R \pm C + LE \quad (2)$$

where Q_{abs} is the radiation absorbed by the leaf and, at thermal equilibrium, R, C and LE are the heat dissipation terms: R the outgoing long wave radiation, C the sensible heat exchange and LE the latent heat lost in evaporating water (transpiration). It follows that if LE is reduced, C and R must increase for equation (2) to remain balanced. Under most conditions, where C and R are sufficiently large, this can be accomplished without too large an increase of leaf temperature (T_1); (that is, $T_1 - T_a < 3^oC$ where T_a is air temperature). Consequently it may be concluded that transpiration is not essential for cooling leaves. This and other possible effects of transpiration reduction have been discussed by Poljakoff-Mayber and Gale.[7] They concluded that only a small fraction of the transpiration, which takes place in normal field crops, is essential, and this mainly for expedition of movement of mineral nutrients from the roots to the plant tops. However removal of surplus heat, to prevent overheating of the plants, is the major problem to be solved in a closed system.

High levels of CO_2 and air humidity are also known to reduce the sensitivity of some plants to salinity (Hoffman and Rawlins[9]; Gale[10]). Hence it may be possible to use the often available brackish water in a closed system. However we calculate that water use will be so small that this may not necessarily be an important consideration in the overall economy of the closed system (see below).

b. Potential Growth

From the point of view of response of plants to supplementary carbon dioxide, two physiological groups (generally termed C_3 and C_4) may be distinguished[8]. As shown in Fig. 1, plants of the 'C_4' type do not show light saturation of photosynthesis even at maximum solar radiation flux densities; neither do they respond to increase of (CO_2) from the normal level of about 320 $\mu \ell . \ell^{-1}$.

Individual plant leaves of the 'C_3' physiological group, to which essentially all the crop plants which are grown in greenhouses belong, achieve light saturation at about one fifth of maximum sunlight. However their level of photosynthesis can be raised by increasing the air content of CO_2. Elevated CO_2 increases net-photosynthesis of these plants, even under low light conditions, especially when leaf temperature is high. It is to be expected that

CO_2 supplementation will increase photosynthesis and hence crop growth, even more under the high light, high temperature conditions of arid regions. A level of 1000 $\mu\ell.\ell^{-1}$

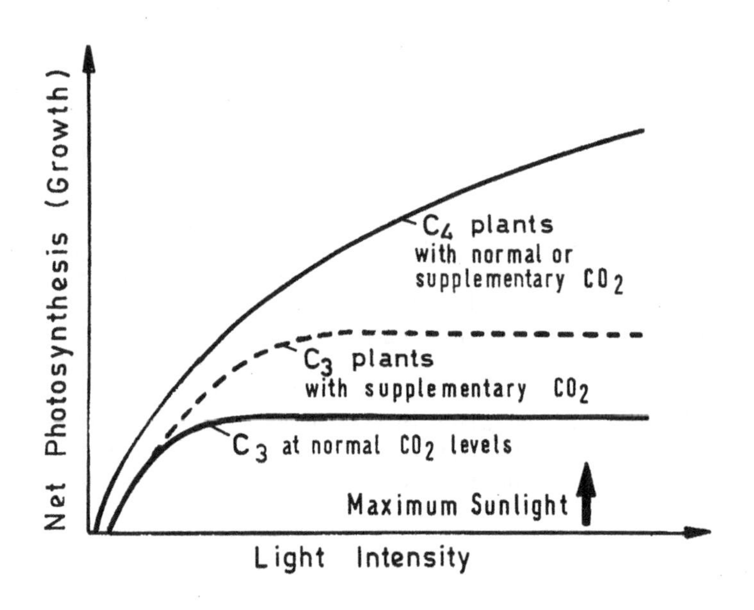

FIGURE 1, Response of 'C_3' and 'C_4' plants to radiation and to supplementary carbon dioxide.

can be expected to raise plant growth and yields by 20-100% (Enoch and Hurd)[11]. The actual response to CO_2 depends upon the environmental conditions and the part of the plant harvested. However, CO_2 is little used in greenhouses in arid regions. The reason for this is overheating. In hot arid regions conventional greenhouses must be ventilated during the very hours of the day when CO_2 supplementation would be most advantageous. Sometimes pad and fan type cooling systems are used which also do not allow for CO_2 supplementation (and are costly in water and energy).

Bassham[2] has analysed the potential for plant growth in closed systems in arid zones, using CO_2 supplementation. He estimates that with a continuously harvested alfalfa crop a theoretical maximum of 200 metric tons/hectare could be produced annually. He proposed that appropriate closed system and protein extraction technology could provide the means for producing protein for cattle, poultry and humans in desert regions.

Extremely high levels of starch production from sweet potatoes grown from single leaf and node cuttings have been reported by Yabuki and Uewada[12]. They grew these cuttings in a closed system with a CO_2 concentration up to 2,400 $\mu\ell.\ell^{-1}$. Starch production of 225 mg dry weight/day/leaf was achieved.

Such high levels of productivity as predicted by Bassham[2], by de Bivort *et al.*,[13] and by Yabuki and Uewada[12] are based on a 12 month growing season of controlled temperatures, optimal root nutrition and water balance, high levels of carbon dioxide and high levels of solar radiation. However, as noted by Yabuki and Uewada[12] and found in greenhouse practice, not all plants respond well to CO_2 supplementation.

III. THE OPTICAL LIQUID-FILTER SOLUTION FOR CLOSED SYSTEMS

It may be said, in a way analogous to the desalination of sea water, that there is no outstandingly difficult scientific or engineering problem in constructing a controlled climate enclosure. The problem is to devise a system which can be built and operated at a price which the farmer can afford.

At Sede Boqer a research team was set up to study the different engineering solutions which have been suggested, which at the start, included economists. After initial study of various engineering solutions, most attention is now being paid to the liquid filter type greenhouse.

Canham[14] described a greenhouse which was cooled by a pigmented stream of water flowing over the glass roof. The liquid absorbed part of the solar energy and could be recirculated at night to reduce cooling. Ideally the liquid filter would transmit the photosynthetically active 400-700 nm waveband of the solar spectrum (48% of the total energy) and would absorb the remaining 200-400 and 700-3000 nm wavebands (52% of the total energy).

Chiapale *et al.*,[15] advanced the concept of Canham, passing the liquid through a 12 mm thick plastic roof and using 2% copper chloride solution as the liquid filter. While having optical characteristics superior to those of the green pigment used by Canham (Solivap green) $CuCl_2$ still absorbs about 30% of the 400-700 nm waveband and, at the optical density used, does not absorb all the 700-3000 nm, near infra-red waveband. Furthermore, copper ion may permanently poison the soil for plants, should the system leak.

We are designing a system, similar in principle to the above, but incorporating an improved filter. The filter absorbs less of the photosynthetically active radiation and more of the near infra-red and is of low phytotoxicity. Furthermore, it has sufficient optical absorbance to allow for a lightweight, plastic, 6 mm thick roof. This system will also include CO_2 supplementation and will utilise hydroponics.

As can be seen by reference to Figure 2, the closed system should operate as follows. During the day, the liquid filter is pumped through the roof, entering at about 18°C and leaving at about 28°C. Some 50% of the solar radiation (mainly the non-photosynthetically active 700-3000 nm waveband) is absorbed by the filter.

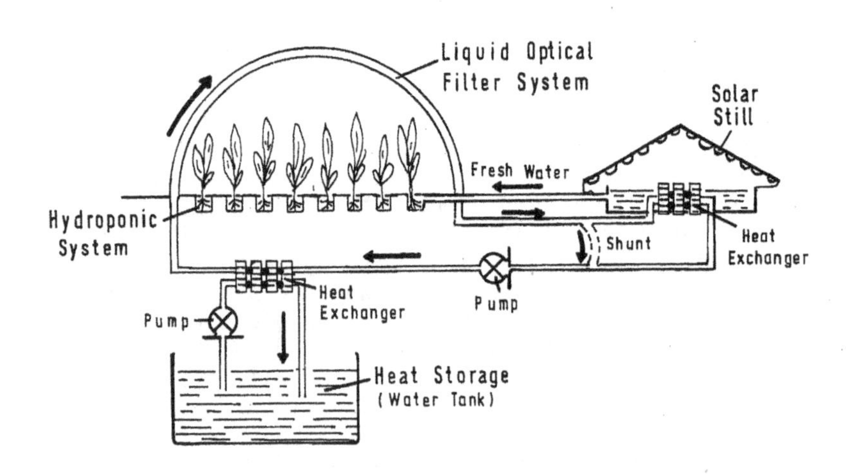

FIGURE 2, Scheme for closed system agriculture in desert regions.

Most of the other 50% of the solar energy, which enters the system, is eventually converted to heat. Only about 4% is absorbed by the plant (in the process of photosynthesis). This heat is transferred from the plants (whose leaf temperatures should not be above 30°C) to the roof, by convection (induced by blowers) and by condensation on the roof. This condensation is important as it a) allows for a slow rate of transpiration, b) transfers surplus heat from plants to roof and c) prevents condensation on the leaves (a most important factor in prevention of disease). This condensate can be recycled back to the roots, via the hydroponics system. A disadvantage of the condensation is a small reduction in transmitted photosynthetic radiation.

The reverse process occurs during the night. The warm liquid of the filter is pumped through the roof where it should cool back to about 18°. This warm roof prevents radiative cooling of the plants and formation of condensate on the leaves.

Our initial calculations show that there is a major problem with this type of system:

Due to the relatively small allowable temperature differential (about 10°C) a very large heat storage is required-about 400 ℓ per sq. meter of greenhouse. A water storage tank of this size would not be prohibitively expensive. However, if this tank is to be filled with the filter fluid (e.g. 2% $CuCl_2$) the cost would be very great. The theoretically most attractive way to overcome this problem would be to incorporate the near infra-red absorbing pigment into the plastic. Experiments in this direction are now under way. A second possibility is to have a heat exchanger between the separated filter fluid and water from the heat storage tank. A third possibility is to reduce the size of the storage by dividing it into a hot and cold tank with a heat pump between.

Yet another possibility is to use eutectic salt mixtures (Telkes)[16] to reduce the size of the heat storage tank. This could be either incorporated directly into the roof (with the optical filter) or used, with a heat exchanger, in the water tank, reducing the required storage volume. However, the reduction in volume is estimated to be only 1:4 to 1:5.

Which combination of the above solutions is optimal from both the engineering and economic viewpoints is at present being studied.

Carbon dioxide levels in the closed system will be maintained at between 1000-2000 $\mu\ell.\ell^{-1}$. Eventually CO_2 levels will be optimised according to growth/light/leaf temperature/profit functions, developed for each particular crop.

IV CLOSED SYSTEM AGRICULTURE AND WATER USE

We calculate that water use in the closed system will be only one fifth to one tenth that of an open field in the same desert region. Furthermore the farmer would need only about 1/20 of the acreage to earn his living. Consequently, we estimate the water saving of closed system agriculture versus open field, conventional, irrigated agriculture, as a factor of 1:100 to 1:200, when calculated on the basis of water use per settled family.

Insofar as a small amount of fresh water will still be required (a completely closed system would be prohibitively expensive) this could be supplied by solar desalination[17] carried out in a shallow, covered-pond type, still (Figure 2).

The conditions prevailing in the arid regions, for which the closed system is planned, are very close to the optimal conditions under which solar desalination may be profitable (e.g. "Appropriate technologies..."[17]). These include: relatively small fresh water requirements, availability of brackish water, plentiful solar energy, inexpensive land area, proximity of supply and demand, absence of necessity for large storage facilities and - a unique advantage of this system - an intrinsic linkage of rate of fresh water production to demand. We calculate that a still of area of about one-fifth of the closed system could provide all the necessary fresh water.

In the final analysis the feasibility of closed system agriculture will depend on many economic factors, such as availability and cost of investment capital, marketability of high cash products and the technological level of the farmers.

V. SUMMARY

Closed systems offer many intrinsic advantages in hot desert environments. Foremost among these are low water use and high yields. Of the total transpiration which occurs under field conditions only a few percent is actually essential for the plant. In arid regions water use in a closed system may be less than one fifth to one tenth that of crops growing in open, irrigated fields. Furthermore, the farmer requires less than one twentieth of the acreage to earn his living. This may result in closed system agriculture using 100-200 time less water than conventional agriculture. In addition, the humid, CO_2 rich atmosphere of the closed system, reduces the sensitivity of some plants to salinity. This may enable the use of brackish water, which is often available, albeit in small quantities.

The second most important potential advantage of closed systems is the possibility of increasing growth and yields. This may be brought about by controlling temperatures, increasing the concentration of CO_2 in the atmosphere and a 12 month growing season. A third major advantage would be the reduction of the use of fossil and conventional energy for heating and cooling, by trapping, storing and recycling solar energy. Closed system grown crops would also benefit from those advantages common to conventional greenhouse crops, such as protection from wind and sand abrasion. Finally, in desert areas, there is the interesting possibility of using solar energy for desalinating brackish or saline water, to fill the relatively small water requirement of the system.

The main problem to be overcome is energy transfer; How to devise an economical method of transferring surplus solar energy (especially radiation of wavelength longer than 700 nm) from day to night. A promising technique for doing this, based on liquid filters, was suggested by Canham in the U.K. A greenhouse operating on the same principle has been built in France by Damagnez and a similar approach is followed at Sede Boqer. Bettaque (Germany) has constructed a combined salt water still and greenhouse at our institute. This we are studying with a view to making it into a closed system.

VI. ACKNOWLEDGEMENT

This work is being carried out in co-operation with scientists of the IBM-Israel Science Centre, the Haifa Technion and the Department of Chemistry Tel Aviv University, and is supported by the KFK Germany.

VII. REFERENCES

1. A. Issar (Ed.) Proc. Int. Symp. Brackish Water (Ben-Gurion University of the Negev, 1975)

2. J.A. Bassham, Science, 197, 630 (1977)

3. Conference on Solar Energy for Heating Greenhouses and Greenhouse Residential Combinations. Ohio Agr. Res. and Development Centre and ERDA, (1977)

4. Proc. Int. Symp. Controlled Environment Agriculture. Environmental Research Laboratory. Univ. Arizona, (1977)

5. H. Meidner and T.A. Mansfield, Physiology of stomata (McGraw Hill, N.Y. & London, 1968)

6. K. Raschke, Planta 48, 200 (1956)

7. A. Poljakoff-Mayber and J. Gale in T.T. Kozlowski (Ed.) Water deficits and plant growth, 3, 277 (Academic Press, N.Y., 1972)

8. I. Zelitch, Photosynthesis photorespiration and plant productivity. (Academic Press, N.Y. 1971)

9. G.J. Hoffman and S.L. Rawlins. Agr. J., 63, 877 (1971)

10. J. Gale in Poljakoff-Mayber and Gale J. (Eds.) Plants in saline environments. (Springer-Verlag, Berlin, 1975)

11. H.Z. Enoch and R.G. Hurd. J. Exp. Bot., 28, 84 (1977)

12. K. Yabuki, T. Uewada, Acta Hort., 87 (1980 in press)

13. L.H. de Bivort, T.B. Taylor and M. Fontes. National Technical Information Service USDC, PB-279-211 (1978)

14. A.E. Canham. British Electr. Allied Indus. Res. Assoc. ERA preliminary report W.T40, (1962)

15. J.P. Chiapale, J. Damagnez, P. Denis, P. Jourdan, 12° Colloque National des Plastiques en Agriculture, Hyeres. Comm. des Plastiques en Agriculture, p. 87 (1976)

16. M. Telkes, Solar Engineering Magazine, (Sept. 1977)

17. Appropriate technologies for semi-arid areas: Wind and solar energy for water supply. German Foundation for International Development, (1975)

Part Two

RESOURCE ECONOMICS OF THE DESERT

RESOURCE ECONOMICS OF THE DESERT

The source of energy for driving the atmospheric circulation is the sun. What happens to this energy, and how it is used to provide resources for desert settlement, is discussed in the following chapters.

Although the general meteorological factors involved in the formation of deserts are known, the details of their size, distribution, and future development are unknown. Indeed, the growth of various deserts of the world in recent years - the desertification process - has sparked much activity in the study of the meteorology of deserts. How to proceed in planning observational and theoretical studies of problems peculiar to the meteorology and climatology of the desert is discussed in the chapter by Louis Berkofsky.

The atmospheric circulation systems peculiar to desert regions are such that rainfall is minimal. It is therefore of utmost importance to manage the rainfall received in an optimal manner. The systems management approach to this problem is discussed in the chapter by Arie Issar.

David Faiman discusses the use of solar energy and its suitability for remote desert regions. He shows how the sun has already been used to reduce fossil fuel consumption in the Negev, and goes on to discuss the technology relevant to the use of solar energy in the planning of future desert settlements.

Uri Regev discusses the economic factors (water development at a reasonable price, technological innovations which would permit the use of solar radiation, support of the infrastructure required as an initial investment) which are required to guarantee that deserts without obvious natural resources be economically viable.

In the discussions of settlement of the desert, an important factor is man's effect on the desert ecosystem. In the last chapter, Moshe Shachak points out the extreme sensitivity of the desert ecosystem because of its unpredictable physical variables (such as rainfall). The unpredictability must not be increased by carelessly introducing the additional factor of man.

DESERT METEOROLOGY

LOUIS BERKOFSKY

I. INTRODUCTION

What causes deserts? The answer to this question is intimately connected with that of the climatic fluctuations within the entire atmosphere. There are both natural and human causes of climatic fluctuations. A list[1] of such causes, along with their characteristic time scales, is shown in Fig. 1. (Autovariation means internal behaviour.)

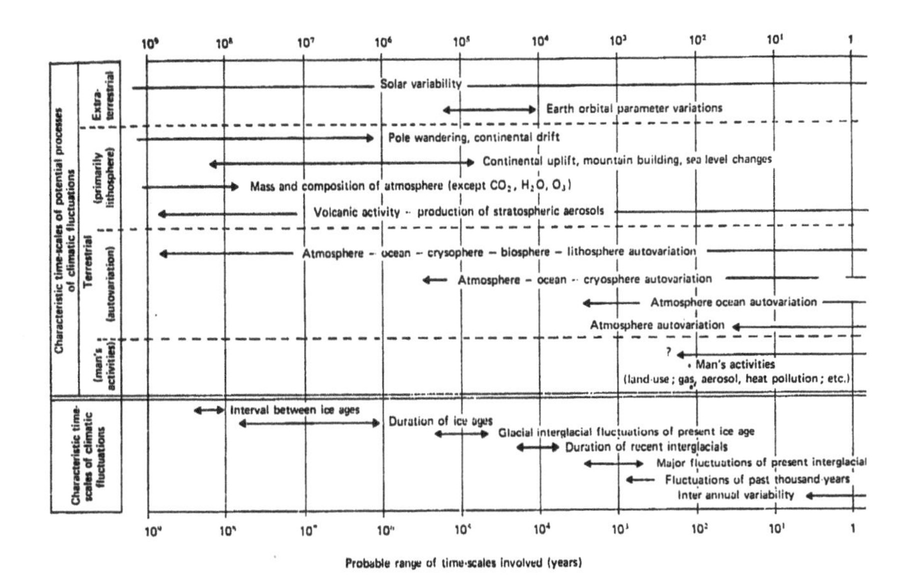

FIGURE 1, Examples of potential processes involved in climatic fluctuations (top) and characteristic time-scales of observed climatic fluctuations (bottom) (Ref. 1).

It is seen that climatic fluctuations may be due to extra-terrestrial causes and terrestrial causes, the latter being broken down into litho-sphere, autovariation and man's activities. The time scales may vary from 1 year to 10 years and more. The time scales for the effect of man's activities on climate are still not clearly known.

Natural terrestrial causes of climatic fluctuation are those associated to a large extent with the atmospheric general circulation. Fig. 2 shows the mean (averaged over latitude circles and over seasons) vertical circulation cells in the general circulation.[2] The southernmost of these cells is referred to as the Hadley cell.

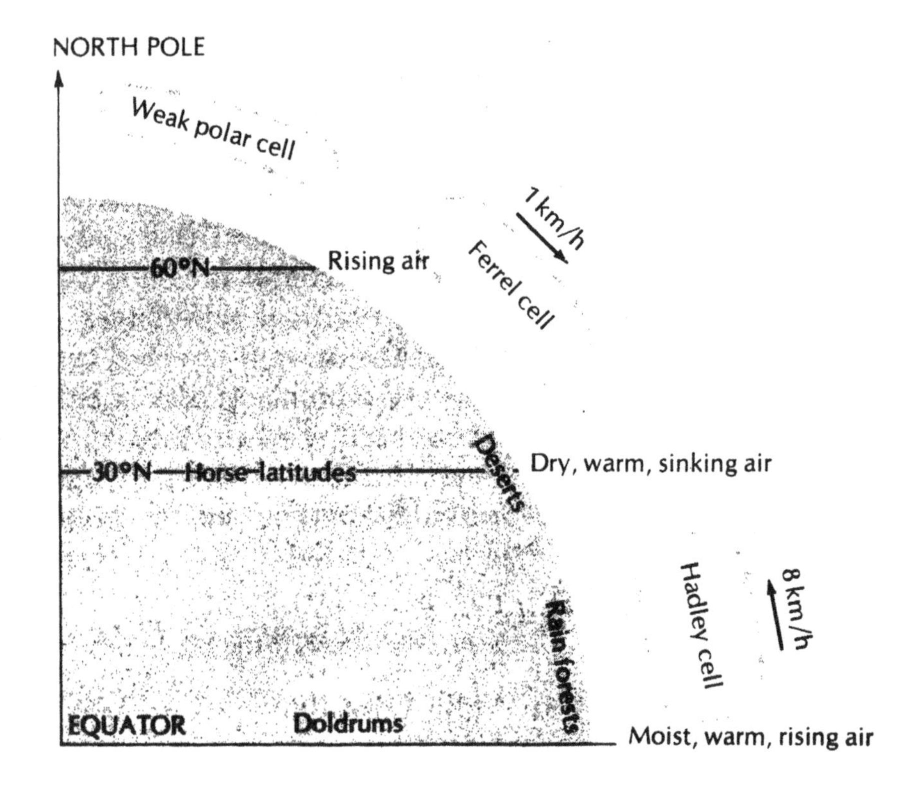

FIGURE 2, Mean vertical circulation cells in the general circulation (Ref. 2).

Fig. 3 shows the mean surface winds associated with the general circulation.[2] Imbedded within the circulation are subtropical high pressure belts, which migrate poleward in summer and equatorward in winter. The subsidence (sinking motion) within these belts warms and dries the air. The mean chart of Fig. 2 indicates the position of the deserts due to sinking. However, this is a mean map, and it should not be inferred that the location of the sinking (and hence the position of the deserts) is homogeneously distributed.

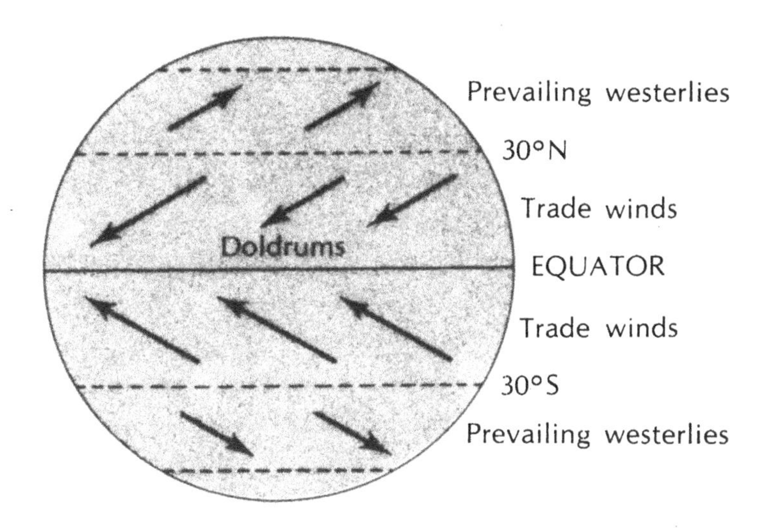

FIGURE 3, Mean surface winds in the general circulation (Ref. 2).

Fig. 4 shows a typical 500 mb (approximately 6 km height) map for a winter day in the northern hemisphere. It is seen that the subtropical highs are well to the south of 30 degrees N. latitude. In addition, there are a number of such cells with relatively lower pressure between them. Thus, there may be interaction between the desert circulation and that farther north and south.

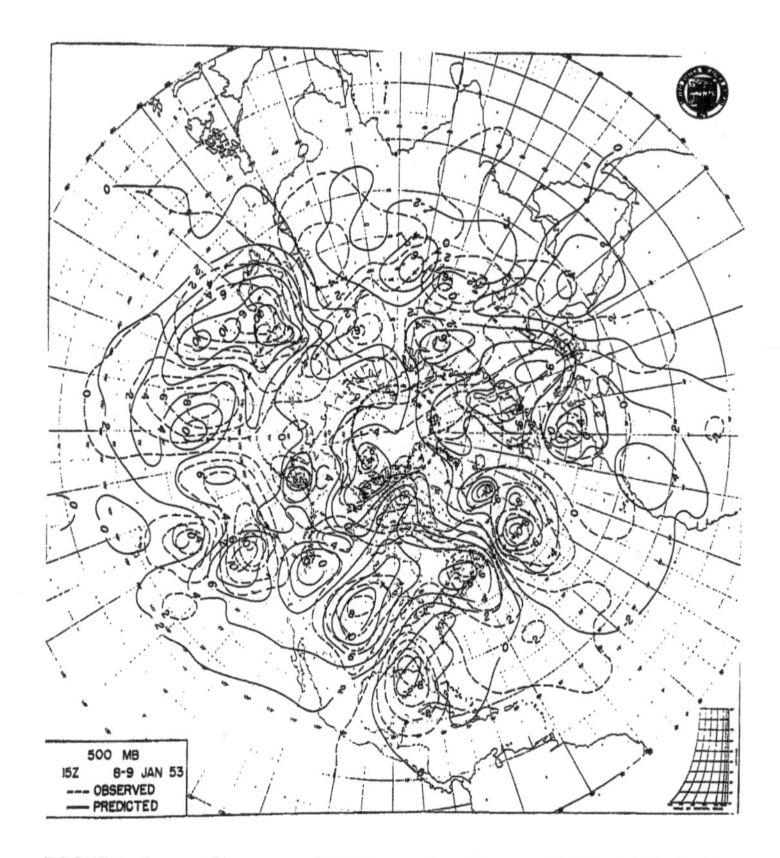

FIGURE 4, Observed 500 - mb flow, 1500 GCT 8 January, 1953. Contours in hundreds of feet.

Fig. 5 shows the rainfall variability.[3] Desert rainfall is characterized by its highly variable nature, both in space and time. It can be seen that many, but not all of the world's arid zones are located at 30 degrees latitude.

Human causes of deserts are those associated with unwise land use practices at times of climatic stress, i.e., during the existence of drought. These practices lead to desertification - the spread of desert-like conditions in arid and semi-arid areas, due to human influence. The dust bowl in the Great Plains of the United States, the degeneration in the Sahel, the Ethiopian plateaux and the Mandoza Province of Argentina are all manifestations of human misuse of the land at times of climatic stress. These are mainly long-range, large-scale effects. Regional and meso-scale effects are also important, however. Examples of regional and local effects due to man-made perturbations are regional deforestation, local or regional thermal, aerosol and gaseous pollution, over-grazing, creation of water bodies, including those used for irrigation, and other agricultural practices.

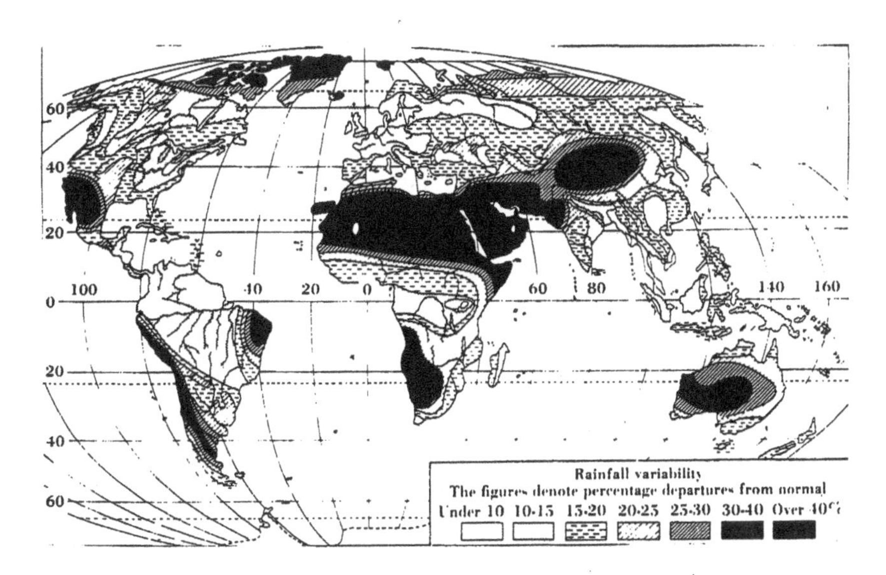

FIGURE 5, The variability of mean annual rainfall for the world (adapted from Ref. 4).

In Fig. 6 we see a possible climatic effect due to clearing of forests.[4] Conceivably, similar practices - in terms of over-grazing - could lead to desertification on a smaller scale. Conceivably, also, creation of water bodies in desert areas could affect the regional desertification process.

How can we investigate climatic fluctuations caused both naturally and anthropogenically? A very powerful approach is to establish a mathematical model based on physical factors. By varying parameters within the model, it becomes possible to study effects of these variations on subsequent climatic changes.

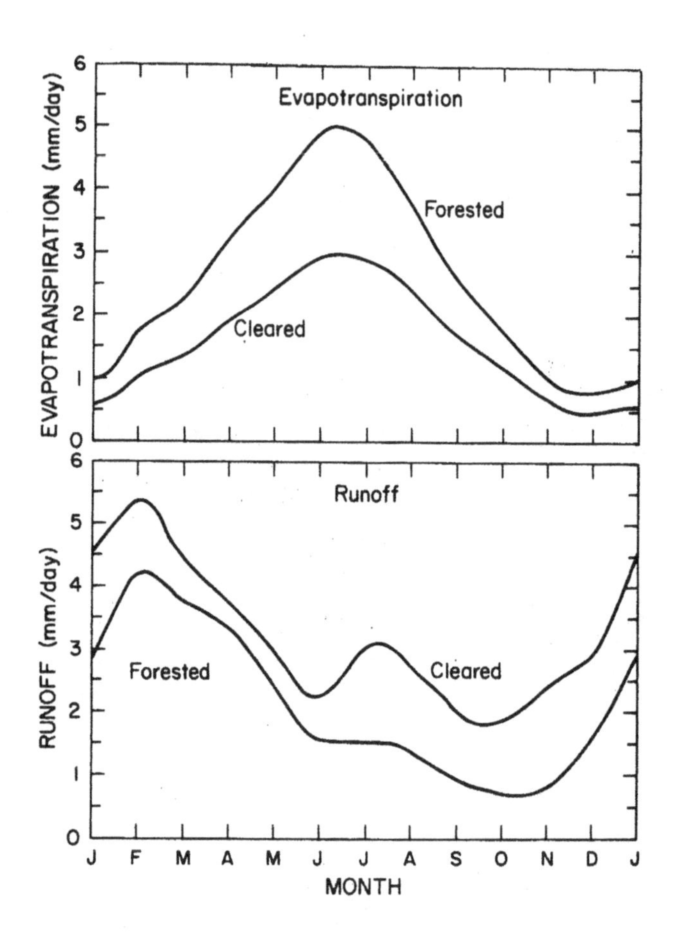

FIGURE 6, (Upper) Evaporation from the land and transpiration from plants (together called evapotranspiration) of water vapour and (lower) runoff of water from forested and cleared land in North Carolina. Clearing land increases the runoff and decreases the evapotranspiration in comparison to forested land in all seasons. Thus, deforestation on a large scale can change the water-bearing characteristics of the land and perhaps even the climate (after Ref. 4).

In Fig. 7 we see the components of the coupled atmosphere-ocean-ice-earth climatic system.[5] The components of the climatic system may be broadly classified as thermal properties, which include the temperature of the air, water, ice and land, kinetic properties, which include the wind and ocean currents, together with the associated vertical motions and the motion of ice masses, aqueous properties, which include the air's moisture or humidity, the cloudiness and cloud water content, groundwater, lake levels and the water content of snow and land and sea ice, and static properties, which include the pressure and density of the atmosphere and ocean, the composition of the (dry) air, the oceanic salinity, and the geometric boundaries and physical constants of the system. These variables are interconnected by the various physical processes occurring in the system, such as precipitation and evaporation, radiation, and the transfer of heat and momentum by advection, convection and turbulence.

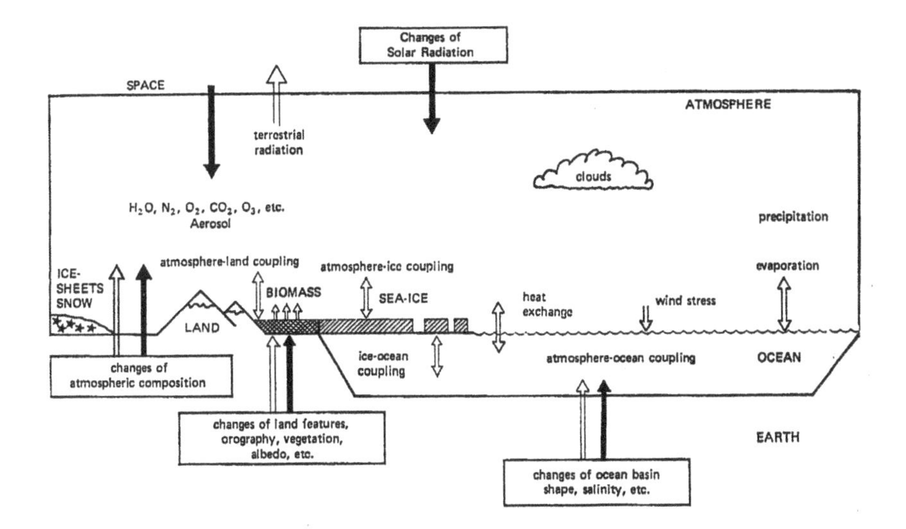

FIGURE 7, Schematic illustration of the components of the coupled atmosphere-ocean-ice-land surface - biomass climatic system. The full arrows (➡) are examples of external processes and the open arrows (⇒) are examples of internal processes in climatic change (adapted from Report of the Panel of Climatic Variation to the U.S. GARP Committee, 1974).

The heating rate is itself highly dependent on the distribution of the temperature and moisture in the atmosphere and owes much to the release of the latent heat of condensation during the formation of clouds and to the subsequent difference of the clouds on the solar and terrestrial radiation. These processes, together with others that contribute to the overall heat balance of the atmosphere are shown in Fig. 8, in which data derived from satellite observations have been incorporated.[6] Here the presence of clouds, water vapour and carbon dioxide (CO_2) is seen to account for over 90 percent of the long-wave radiation leaving the earth-ocean-atmosphere system. This effective blocking of the radiation emitted by the earth's surface, commonly referred to as the greenhouse effect, permits a somewhat higher surface temperature than would otherwise be the case. It is interesting that this important effect is achieved by gases in the atmosphere that exist in near trace amounts.

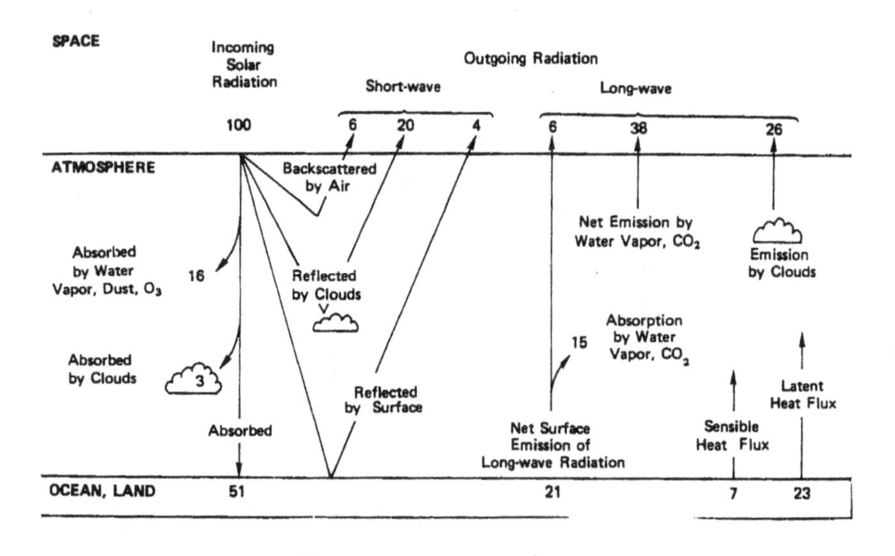

FIGURE 8, The mean annual radiation and heat balance of the atmosphere, relative to 100 units of incoming solar radiation. Based on satellite measurements and conventional observations (Ref. 6).

We see from Fig. 8 that the role played by clouds is an important one: the reflection and emission from clouds accounts for about 46 percent of the total radiation leaving the atmosphere; and in terms of the shortwave radiation alone, clouds account for two-thirds of the planetary albedo. The largest single heat source for the atmosphere is that supplied by the release of the latent heat of condensation, and this is particularly important in the lower latitudes. There is also an appreciable supply of sensible heat from the oceans, especially in the middle and higher latitudes. It is therefore clear that water substance, in either vapour or droplet form, plays a dominant role in the atmospheric heat balance. And when we recall that the oceans themselves absorb most of the solar radiation reaching the surface, and that the presence of ice and snow also affect the heat balance, the climatic dominance of global water substance becomes overwhelming, even if ice is not taken into account.

The above description of the global radiation balance stresses the importance of water. The desert radiation balance is far less sensitive to water. Yet, to understand the desert circulation itself, and the influence of the desert circulation on the global circulation, the details of the desert radiation balance must be known. These details will provide important input into the energy equations, which form a major part of mathematical systems for studying climatic change.

A great deal of thought has been given to the possibility of modifying the desert climate, thus reversing the desertification process. In order to think in these terms, one must be able to observe, understand and predict before one can modify the atmospheric circulation. It must be admitted, however, that modification both advertent and inadvertent, has been brought about at times without a complete understanding of the phenomena being modified.

II. A DESERT METEOROLOGY UNIT

A Desert Meteorology Unit should concern itself not only with the causes of deserts, but also how any information obtained may be used as an aid in desert settlement and agriculture. Thus its research programme should essentially be two-pronged: to observe appropriate parameters, and to develop mathematical models making use of these observations. The observations themselves serve as a basis for understanding the desert atmosphere, and the mathematical models then serve as a basis for prediction and modification.

a, Observations

What should be observed? Heavy emphasis should be placed on boundary layer (height 0-1 km) observations, since processes taking place there have a strong influence on the entire desert atmospheric circulation, as well as being of importance in themselves for desert settlement and agriculture. Such observations should include wind, temperature and humidity profile measurements, components of radiation, as well as inversion height, cloud condensation nuclei and dust. The thermo-dynamic variables just mentioned form the basis for a mathematical model of the climate near the ground. A fuller description will be given later.

Observations of cloud properties, that is to say, their dimensions, growth rates, liquid water content, rainfall rates, should be taken with a view towards understanding the structure of clouds. These observations will form a basis for predicting cloud life cycles, and for approaching the problem of increasing desert rainfall my modifying the clouds. It may be worth-while at this point to discuss the rainfall augmentation problem.

Since there are relatively few desert clouds, and there is a relatively small amount of desert rainfall, desert rainfall augmentation experiments must be designed to yield positive results. It does not seem worth-while to increase the annual rainfall by 10-15% in an area where the annual rainfall is normally 100 mm. But if the following view is taken, augmentation experiments may be worth-while: desert rainfall is very spotty, and the precipitation is frequently heavy and results in small desert floods. By a method known as "dynamic seeding" i.e. dropping a "bomb" of silver iodide into a cloud top during the initial rain stage, overseeding, releasing large amounts of latent heat of condensation, thus increasing the buoyancy, and causing large and rapid vertical growth of the cloud, it has been possible to double the rainfall from an individual cloud. Thus additional desert floods may be caused. Since desert floods are very beneficial for desert agriculture, dynamic seeding may be a potent factor in helping to settle the desert.

Observations of the wind, pressure, temperature and humidity in the free atmosphere ambient to clouds are necessary for studying cloud growth rates.

Observations of rainfall, both by ground-based and remote sensing equipment are important both for information purposes and for planning modification experiments.

Finally the components of turbulence, and their vertical variation, should be measured. Until the characteristics of turbulence are understood, there is little hope of predicting small-scale motions in the atmosphere. It should be emphasized that wherever possible, full use should be made of satellite observations.

b. Prediction

What mathematical models should be developed and for what purpose should they be used? There are a wide variety of models of the large-scale atmospheric circulation. Such models may be able to predict certain aspects of desert evolution, but smaller-scale models both in space and time are needed to study finer details of the evolution of the desert circulation. It follows that it is worth-while for a desert meteorology unit to develop meso-scale and micro-scale models, with the possibility to imbed them within large-scale models. Then boundary conditions for the smaller-scale models may be obtained from predictions of the larger-scale models.

While the possibilities are limitless, we describe below several examples of desert circulation models which may be applied to a wide variety of problems.

1. A meso-scale boundary layer model to investigate anthropogenic effects.

It is possible to develop a mathematical model of the planetary boundary layer (depth approximately 1 km) which not only has the capability of predicting the evolution of the low-level circulation, but may also be used to evaluate the effect of an artificial body of water. In Fig. 9, we see schematically how an early morning inversion is modified by convection and turbulent mixing during the day. When stably stratified warm dry air flows from land over cooler water, the warm air is cooled due to a downward transfer of heat to the cooler water. At the same time, turbulent mixing takes place, destabilizing the profile into that shown schematically in Fig. 10. Notice that in both cases, the inversion height changes with time, so that this variable must be predicted. Indeed, since most deserts are near the edges of Hadley cells, the existence of inversions within desert air is so common that their presence cannot be ignored. Similar schematics to those in Figs. 9 and 10 exist for the moisture distribution.

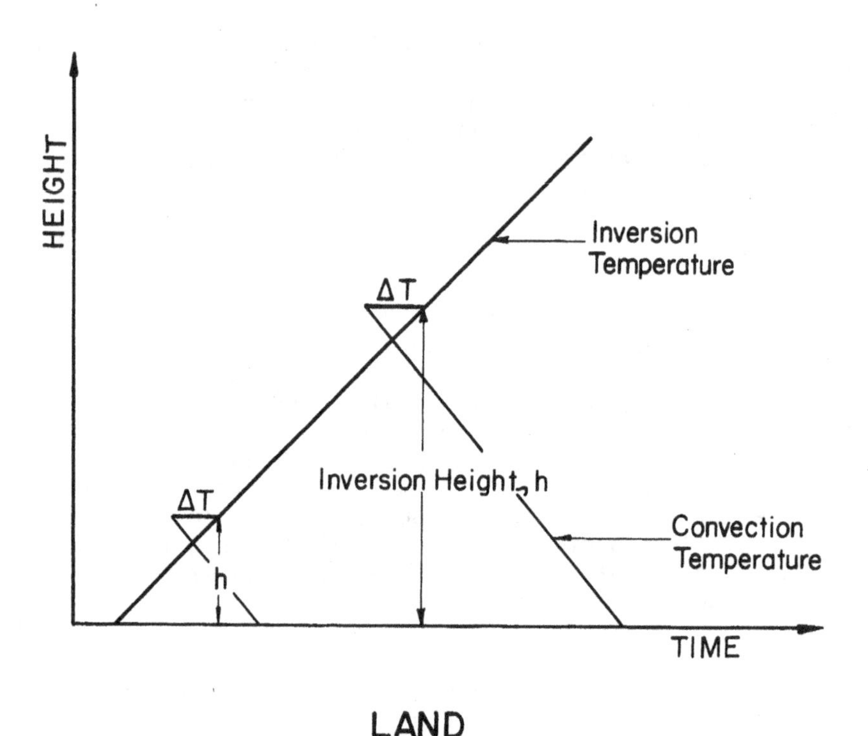

FIGURE 9, The modification of air temperature over land from early morning (inversion temperature) to mid-afternoon (convection temperature).

Additionally, in this model the net radiation depends upon the ground temperature, which in turn depends upon the soil moisture. Therefore, these variables must also be predicted.

Studies of the influence of mesoscale features (irrigation projects, desert regions, patches of forest, cities, etc.) by means of instrumented aircraft have shown that strong heat transport caused high evaporation from a small lake, with a cool layer which extended well beyond the lee of the lake.[7] The flux of water vapour over irrigated land was essentially double that over surrounding non-irrigated areas. A small dry city produced a heat island which delayed development of a temperature inversion for up to nine hours.

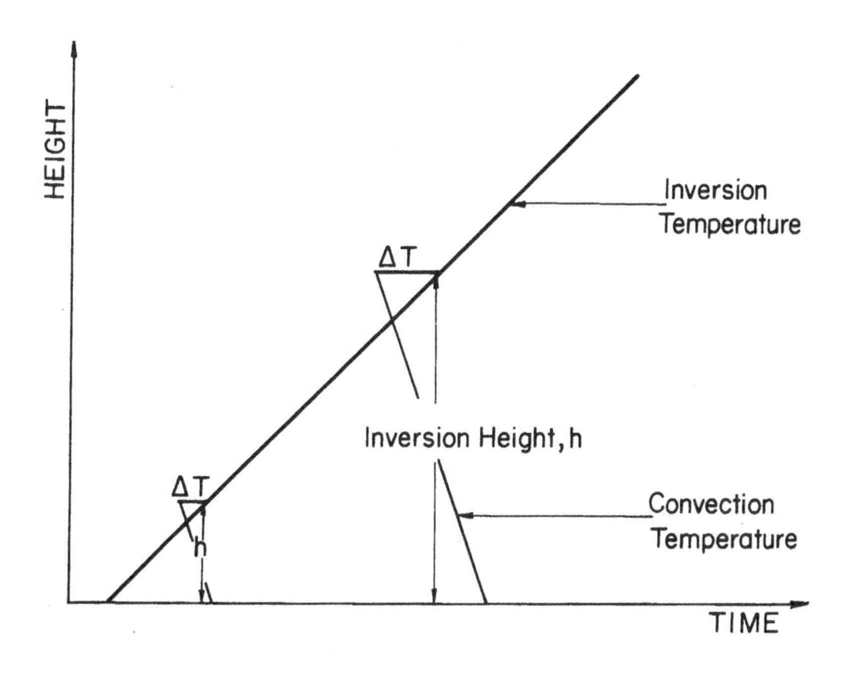

WATER

FIGURE 10, The modification of air temperature over water from early morning (inversion temperature) to mid-afternoon (convection temperature).

The model described above is capable of simulating such effects. Indeed, if the construction of artificial bodies of water is contemplated, their micro-meteorological effect may be studied in the following way: establish an observational network in the region of the proposed body of water. Test the validity of the model by making predictions within the observational network region. Replace several points within the network by a simulated body of water, predict again, and attribute any differences to the presence of water. This procedure enables one to determine what size body of water will exercise some control over the surrounding climate.

2. A mathematical model for predicting the climate near the ground.

The basic concept used in the development of the model is the existence of an energy balance at the earth-atmosphere interface at any time. This can be written $R_N = H + LE + S$, where R_N is the net radiation, H is the turbulent heat transport, E is the evaporation rate, L the latent heat of water and S the flux of heat into the soil. If the net radiation is given as a function of time, we can write expressions for H, LE and S in terms of temperature gradients above and below the earth's surface, and in terms of humidity gradients above the earth's surface. The expressions for the atmospheric fluxes assume a basic logarithmic wind profile near the ground modified by a stability parameter. In addition, the fluxes are assumed to have a given vertical profile. On the other hand, the flux of heat into the ground is solved directly by incorporating the time-dependent conduction equation.

By specifying the wind, temperature and humidity at the upper boundary (about 200-300 m above the ground) and the temperature in the deep soil, it is possible to simulate the "climate" near the ground as a function of time, and to study the effect of soil parameters and albedo on it. In particular this model has been used to simulate dew formation on the dry soil and its evaporation after sunrise to study the effect of dew on the energy budget near the ground.

3. Mathematical models for studying albedo circulation interaction.

Due to the recent drought in the Sahel region of West Africa, there has emerged a great deal of discussion concerning the causes of such droughts. Some attribute drought to desertification caused by poor land management[8] (although, as stated earlier, poor land management during drought leads to desertification). Some say that surfaces denuded by overgrazing reflect more radiation than those in which dead litter is distributed, are therefore cooler than their surroundings under sunlit conditions, and there is decreased lifting of air necessary for cloud formation and precipitation.[9] Some suggest that the high albedo of a desert contributes to a net radiative loss relative to its surroundings, and that the resulting horizontal temperature gradients induce a frictionally controlled circulation which imports heat aloft and maintains thermal equilibrium through sinking motion and adiabatic compression. That is, radiation and friction produce strong sinking motion over the (high albedo) desert.[10] In practice, albedo is changed by removal of vegetation by drought, overstocking, cultivation, or dessication of the soil itself, soil albedo being related to soil water content. These three mechanisms are likely to occur simultaneously.

The above description is that of a large-scale model. A simplified version of the Meso-Scale Boundary Layer Model may be used to test the hypothesis that strong sinking motion is related to high albedo,[11] even on the meso-scale. Table 1 shows how the velocity decreases as albedo increases. A time integration of this model,[12] Fig. 11, showed that the effect of a uniform decrease of ground albedo was to increase the maximum value of the upward velocity and to decrease the corresponding values for the downward velocity. However, a more dramatic effect was the impression of an albedo gradient, as is seen from the results for albedo =0.15 at the middle 3 points and 0.35 elsewhere.

TABLE 1. Albedo vs. vertical velocity, northern Negev, Winter

Albedo	Vertical Velocity (Negative value means upward velocity)
1.0	0.77
0.9	0.73
0.8	0.56
0.7	0.47
0.6	0.36
0.5	0.26
0.4	0.16
0.3	0.06
0.2	-0.04
0.1	-0.14
0.0	-0.24

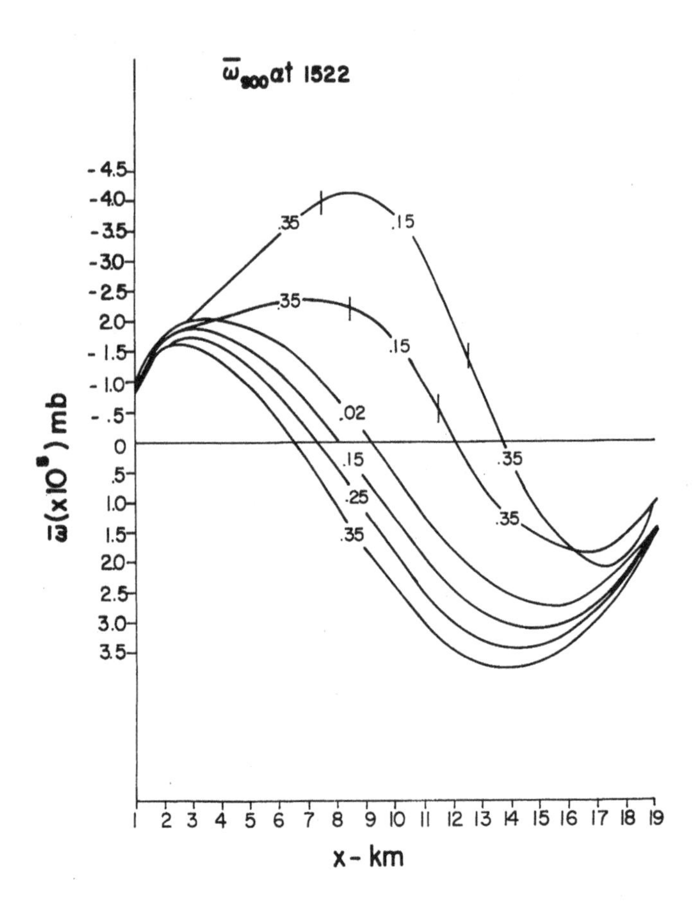

FIGURE 11, Vertical velocity at a fixed time for various values of albedo (Ref. 12). Negative values mean upward motion.

4. A mathematical model for prediction of formation, growth and dissipation of a cloud

The problem of the formation, growth, and dissipation of a cumulus cloud is so complex that it has so far defied solution. The cumulus cloud is a three-dimensional space phenomenon, and should be treated as such. During the last few years, several three-dimensional convection models have been constructed.[13] The number of models has not been great, mainly because they demand cumbersomely large amounts of computing storage and expenditures if adequate spatial resolution is to be achieved. Indeed, even in the most sophisticated model, the calculations had to be limited to preliminary 'midget' experiments with sufficiently few grid-points to circumvent these demands.

By contrast, in recent years, a number of two-dimensional models have been developed for the study of precipitating convective clouds.[14] Two-dimensional axisymmetric models have also been developed.[15]These models demand far less computer time than three-dimensional models, and are able to reproduce many important features of the cumulus cloud. Even simpler, less time consuming models are 'quasi-two-dimensional' models.[16] Such models have given remarkably reasonable results in cumulus dynamics studies, especially when used for operational purposes. Therefore, it seems reasonable to develop a 'quasi-two-dimensional' model for desert cumulus work.

In this model, the variables are functions of height and time. The model is axisymmetric, but the radius of the cloud is itself allowed to vary with height. Thus the cloud model has the characteristics of a 'plume' model. Equations for the entrainment rate, the cloud temperature, the cloud saturation mixing ratio, the cloud water, the hydrometeor water, the fall velocity of water drops, the vertical velocity and the cloud radius are solved simultaneously to yield cloud parameters as functions of time. Input data, such as cloud base radius, temperature and humidity of ambient air, come from the observational programme.

The results of cloud observations, rainfall, and modelling studies may be used to determine whether it is possible to seed desert clouds.

c. Modification

We have concerned ourselves thus far with the problems of observation, understanding and prediction. What can be said about modification? We have described models by means of which we may be able to simulate the effects of changing albedo on the meso-scale climate, the effects of artificial bodies of water on the meso-scale and micro-scale climate. The results of these mathematical simulations will indicate to what extent meso- and micro-climates may be modified. For example, over how small an area may we alter the albedo and still expect a climatic effect? What is the minimum size man-made water body which will yield climatic effects? What is the time scale of such effects? Is it possible or even worth-while to attempt reversal of the desertification process by large-scale plantings? How much rainfall increase might be expected from 'dynamic' seeding of individual clouds? What is the cost/benefit ratio of the above efforts?

None of these questions can be answered without a comprehensive, carefully planned research programme.

III CONCLUSIONS

The importance of the deserts for the atmospheric circulation has been amply brought out in recent years. In recognition of this importance, increasing emphasis has been placed on studies of desert atmospheric circulations by mathematical models. An important ingredient, without which meaningful numerical solutions to the mathematical model equations cannot be obtained, is the initial state of the desert atmosphere. That is to say, meteorological desert data are needed 3-dimensionally. Yet data of this type are sadly lacking. The primary reason for this state of affairs is that the deserts are sparsely populated, and, even where settlements exist, the economic and educational situation prohibits the taking of necessary scientific observations.

This situation can be alleviated, at least partially, by intensive observational and theoretical studies by qualified personnel in those desert regions which are habitable. The observations should be ground-based, and also remotely sensed, using, to the extent possible, upper air instruments carried aloft by balloons and aircraft and artificial satellites.

These are the directions which are being taken in the Meteorology Unit of the Institute for Desert Research.

IV. REFERENCES

1. J.E. Kutzbach, WHO Bulletin, 23, 47 (1974)

2. R.A. Anthes, H.A. Panofsky and J.C. Cahir, The Atmosphere (Charles E. Merrill, Ohio, 1978), pp 110,111.

3. S. Petterssen, Introduction to Meteorology (Mc Graw-Hill, New York, 1969), 338 pp.

4. W.D. Sellers, Physical Climatology (University of Chicago Chicago Press, Chicago, 1965). p. 129

5. U.S. National Academy of Sciences, Understanding Climatic Change, A Program for Action (National Academy of Sciences, Washington, D.C., 1975), p. 14.

6. T.H. Vonder Haar and V.E. Suomi, J. Atmos. Sci. 28, 305-314 (1971)

7. R.M. Holmes and J.L. Wright, J. Applied Meteor. 17, 1163-1170 (1978)

8. N. Wade, Science, 185, 234-237 (1974)

9. J. Otterman, Science, 186, 531-533 (1974).

10. J. Charney, Quart. J. Roy. Meteor. Soc. 101, 193-202 (1975).

11. L. Berkofsky, J. Applied Meteor, 15, 1139-1144 (1976).

12. L. Berkofsky, Beiträge zur Physik der Atmosphäre, 50, 312-320 (1977).

13. R.E. Schlesinger, J. Atmos. Sci. 30, 879-893 (1975).

14. T. Takahashi, PAGEOPH, 113, 950-969 (1975).

15. S. Soong and Y. Ogura, J. Atmos. Sci. 30, 414-435 (1973).

16. A. Weinstein and L. Davis, NSF GA-777, 43 pp (1968)

THE ROLE OF NON-REPLENISHABLE AQUIFERS IN DEVELOPMENT PROJECTS IN ARID REGIONS

ARIE ISSAR

I. INTRODUCTION

The basic problem of life in arid zones is the randomness in time, volume and area of precipitation. Hence there is a scarcity of runoff, ground water recharge, and springs, and thus a scarcity of vegetation and life forms. Therefore, living conditions and frequently survival in these regions depend on the provision of water storage to answer the need when the annual supply is below the minimum required for existence. The magnitude of storage is of decisive importance when the annual minimum requirement is not met for several successive years. In fact, basic survival in many arid countries during negative, fluctuating climatic conditions, then depends on storage. Different communities in history have solved their storage problems in arid zones in different ways. The simplest solution is nomadism, where storage is supplied by space. More elaborate systems involved storing surface water that had been provided by rivers from more humid regions, either by building dams or canal systems. Some communities developed methods for tapping ground water. Another way to ensure existence was by storing food or capital. Certain deserts were barriers between producers and consumers of food or spices and the inhabitants had to trade or tax trade. The reliance on residential border communities to store capital and their persuasion, sometimes by aggressive means, to share this capital with their nomadic neighbors was also a method of survival used during periods of deficiency, and as a sport during periods of abundance.

II. STORAGE

Focusing our discussion on methods which involve the storage of water resources, it is important to discuss the problems of storage from the hydrological (i.e. availability of supply) as well as the hydrogeological (i.e. availability of space) point of view. We can define storage capacity (which involves available water as well as space) as the volume of water exceeding the algebraic sum of period utilization and loss that is capable of being stored for future periods.

Thus, if in a particular region we deduct from the total annual amount of precipitation the total amount of evapotranspiration and compare it with the total annual community requirements, the difference, whether positive or negative has to be added to or subtracted from the storage supply. Whether the positive peaks are able to provide for the negative troughs depends on the availability of storage capacity. Since water is a fluid, storage means working against gravity and evaporation. These two factors work against each other; a high ratio of water surface to volume of storage has a positive significance from the point of view of the economics of the storage structure, but negative from the evaporation standpoint. Moreover, the sporadic supply makes surface storage economically more problematic as it involves considering random floods of low probability but high intensity.

The storage of surface water also has its drawbacks from the general point of view of resource economics. Various observations in arid zones (Evenari, et. al.,)[1] have shown that the ratio of runoff to precipitation decreases as one goes from small to larger watersheds (Fig. 1). These observations require further research, as the influence of local rain cells and the significance of ground water recharge were not thoroughly investigated. In any case, when considering the problem of water supply to small watersheds in developing countries, the efficiency of small structures seems to be higher than large ones because the latter involve water transportation. But even in the arid zones of developed countries, storage of surgace water on a large scale is problematical, and in many instances the crucial problem of long-term shortages has not been solved. One should therefore look for another type of storage which may help in evening out the fluctuations between the extremes of supply and demand on a long-term basis.

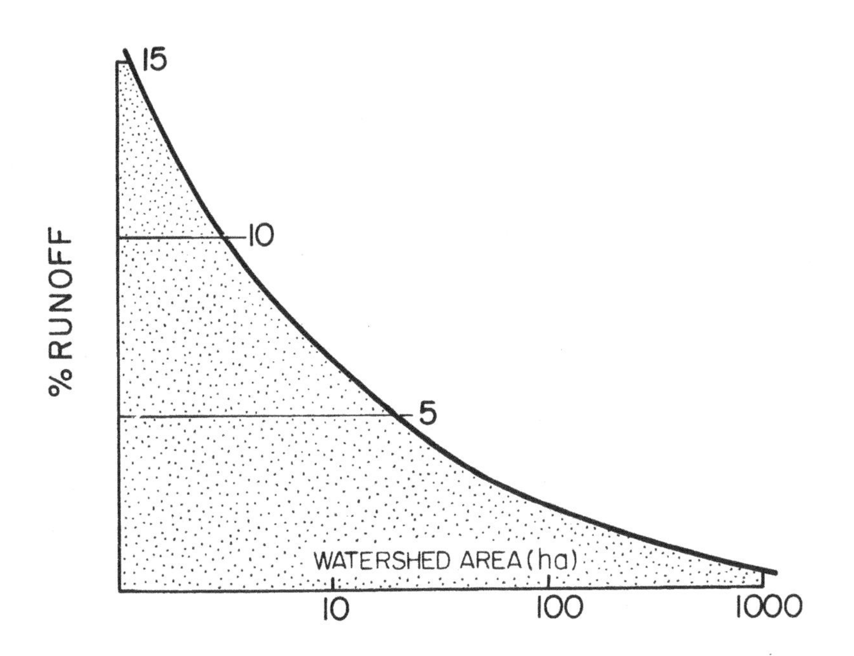

FIGURE 1, Watershed area - runoff ratio.

a. Underground Storage

The optimal storage alternative seems to be ground water, as it can be regarded as storage by a retardation of flow. Thus, every drop of water in the aquifer is forced to go a much longer way due to billions of minute retarding dams. In other words, the effect of the gravity field is reduced through easing the flow gradient, by inducing it to flow through porous media, and the effect of evaporation is minimized by insulation from the atmosphere.

The fact that the water is stored in porous media also has its drawbacks, as part of the volume that would have been available for storage is instead occupied by rock or soil material. The efficiency of rock as storage is defined by its storativity coefficient (S) which gives the percentage of water available versus the volume of aquifer drained. For confined aquifers, it gives the amount of water gained to the amount of pressure reduction.

The efficiency of the rock as a transmitting medium is defined by its permeability (K) or transmissivity (T) which defines the quantity of water flowing through a particular section during a certain period.

Thus, on the one hand, we would like the permeability to be high enough in order to increase the efficiency of recharge and pumpage. On the other hand, the permeability should not be so high as to create a high rate of outflow. For example, the high transmissivity of the gravel aquifer of the alluvial fans on the coast of eastern Sinai causes a high rate of ground water outflow into the sea, which drastically limits the volume of water that can be stored in these fans.

The optimal natural storage medium is sand and sandstone. Outflow can be reduced either by pumping or by the use of subsurface dams. Subsurface storage can not only be made more efficient, but it can be created, for instance by building a dam and filling it with sand or gravel. The problem of course is economics.

In general, storage of water underground can be classified according to its order of magnitude. One can start from storage in the unsaturated zone, namely seasonal soil storage, or field capacity for field crops. (Plowing is the most ancient method of inducing recharge as well as providing storage space by increasing the soil permeability; its antiquity is verified by the common linguistic root of plow and "Falach" which in Semitic languages means either "plowing" or to "cut through".)

A higher degree of storage capacity is afforded by the root zone. The negative aspect of storage in the unsaturated zone and the root zone is the limited volume of storage involved, since it is limited to the upper part of the soil and is under the seasonal influence of evapotranspiration.

Moving up the scale of storage capacity and collection areas, we come to small catchment areas where water harvesting methods, and small dams or water tanks are involved. Storage can be increased by transporting the accumulated water to an area where it can recharge the subsurface or by covering it with some material that reduces evaporation such as plates or monolecular layers. In most cases, this storage is annual or lasts for only a few years.

The possibility of using small dams to form subsurface lenses of ground water along riverbeds and using the water stored in the subsurface either by constructing subsurface drains in the riverbed or by planting trees, the roots of which can benefit directly from the stored water, is another aspect of this type of storage.

b. Aquifers

Aquifers in consolidated or unconsolidated rocks are higher on the storage capacity scale, as due to their larger volume they may be useful for long-term storage. An increase in the storage capacity is thus a function of the hydraulic as well as the geometric properties of the aquifer. The management of the aquifer storage capacity should be a function of a long-range regional development policy, whose objective is the maintenance of, and possibly the increasing of, per capita income of the inhabitants.

This objective can undoubtedly be achieved by other means as well, such as by improvements in crop varieties and advances in agricultural techniques. In this article however, we are concerned with introducing the best system for improving the storage capacity of the scarcest resource in arid regions, namely water.

Aquifers should be viewed from two aspects: (1) renewable storage, and (2) one-time storage or stock reserve. When dealing with the renewable storage capacity of certain aquifers, the first constraints to be taken into consideration are environmental, namely the irreversible negative changes which can effect the aquifer due to pumpage beyond the "safe yield". The second constraints are socio-economical, namely guarantee of the supply on a long-term basis.

This calls for a systems approach for planning the use of storage over time, the objective of which is the optimal deployment of renewable storage (either naturally or artificially induced) during drought periods of high probability. Uncontrollable outflows from the aquifer system should of course be deducted from the gross safe yield of the aquifer. On the other hand, a controlled reduction of the water table may help in limiting these outflows, and the volume gained by the reduction of the water table from its original position should be regarded as a one-time reserve and introduced as such into the system.

In this case, the mined water should be introduced as a finite, limited resource that may help in achieving some initial economic goals, but has to be replaced. This particular position should be viewed separately from the case where a regional aquifer exists with a large stock reserve and storage is used by mining a non-replenishable one-time reserve. The conventional approach is to also regard this resource as an interim supply, anticipating the introduction of other alternative resources which have lower priorities than the one-time reserve. For example, mining of water may be considered before a big dam is built, the construction of which has to be postponed either because of lack of funds or lack of hydrological data or both.

The problem of planning a system which involves water mining with a conventional alternative resource includes separating the project into the following stages: (a) the mining stage, (b) the recovery stage, and (c) the steady stage.

Stage (a) has to include the time required for the collection of data and capital, planning, and execution of the mining of the alternative resource. Altogether, the two resources already at a steady-state in the project have to flatten the curve of availability to an optimal level. Today, such planning procedures can be regarded as conventional and are applied or should be applied in many regional development plans.

c. Non-replenishable Aquifers (Fossil)

The question now arises as to how to approach long-term, one-time reserves when no alternative resource is available or when alternative resources are highly questionable. For instance, there is no predictable, economically feasible technology for the introduction of alternative resources such as sea or saline ground water that require desalination. This is the case in most arid zones of the world where fossil aquifers or "Paleo-water" resources exist.

These aquifers are found under the arid belts extending along ancient margins of ancient geological continental shields. These include the Saharan, Western, and Saudi Arabian Deserts along the northern margins of the African Shield, the Kalahari Desert on its southwestern margin, the Great Artesian Basin east of the Australian Shield, the Central Asian basins south of the Siberian Shield, and Paleozoic and Cretaceous consolidated aquifers in central North America deposited along the southern margins of the Canadian Shield.

A typical basin of this type is the Chad Basin

(UNESCO).[2] This basin, with Lake Chad as its lowest level, constitutes a vast reservoir of subterranean water and is one of the largest closed river basins in the world. The surface area of the drainage basin is 2,335,000 km^2.

The basin is enclosed by a series of crystalline massifs of Precambrian origin, which are outcrop areas of the Basement Complex which comprise the African Shield. For long periods following the Upper Tertiary, the basin was covered by a vast lake which changed its shape according to fluctuations in the climate.

Two main zones of confined aquifers were observed in this basin. These are termed the Lower Zone and the Middle Zone. Another phreatic aquifer called the Upper Zone covers most of the area. The Lower Zone (also called the Continental Terminal and Lower Chad Formation) is of Tertiary age. It was deposited along the edges of the Chad Basin under a wide range of conditions.

From the practical point of view, the water in the confined aquifer may be regarded as *fossil water*. As these layers were deposited under continental, aquatic and terrestrial conditions, they were not later enriched by residual salts. Thus, although fossil, the water in these aquifers is not very highly mineralized.

A phreatic aquifer can be found in some parts of the Upper Zone. In most cases however, the elevation of the water table of these aquifers is lower than that of the piezometric surface of the underlying confined aquifer. If conditions permit (e.g. where an aquitard separates the two aquifers) upward leakage will take place. The areas of such upward flow can be located by tracing water mounds in the water table.

The other basins which extend over vast areas along the margins of the ancient shields arose under continental conditions over long geological periods. The erosion products from the shield were deposited along the margins of these basins in continental or shallow sea environments. These margins were gently folded, if at all, and thus the continuity of the layers was undisturbed. The sea encroached into these basins from time to time in the form of major transgressions. During such periods marls and limestones were mainly deposited. Four main factors caused the formation of regional aquifers in these regions:

(1) The occurence of thick layers of continental deposits of sandstone.

(2) The gentle dip from the shields toward the basins and the lack of severe folding and faulting.

(3) The humid paleoclimates which facilitated the recharge of the aquifers.
(4) Outlets which allowed the aquifers to be flushed.

A typical basin of the kind described above is the Great Artesian Basin of Australia (Hind and Halby; United Nations).[3,4] This basin has an expanse of 1,715,000 km^2 in area and underlies parts of the states of Queensland, New South Wales, South Australia, and the Northern Territories. Three main aquifers are distinguished from bottom to top: the Bundemba Group of Triassic-Jurassic age which is composed of coarse arkosic sand; the Marburg Formation of Jurassic age composed of quartzose sandstone of shales, and Blythesdale Group aquifers which extend over the entire basin. The Bendemba and Marburg aquifers are smaller.

At present, replenishment occurs only along the eastern margin of the basin, i.e. along the slopes of the Great Dividing Range which coincides with the area of highest elevation and highest rainfall. However, prior to the uplifting of these belt areas the main replenishment areas were to the west along the margins of the Australian Shield during the more humid period which prevailed in the late Cretaceous and Tertiary periods. Natural outflows presently occur toward the north or toward the Gulf of Carpentaria; in the past they may have occurred along the eastern margin of the basin. The gradients, although not precisely mapped, show a general trend toward the northeast. Flow rates however, are very small as the head drops a few hundred meters over a flow length of more than a thousand kilometers. This means that replenishment is negligible as compared with utilization and water that is taken from storage. The utilization of water in 1968 amounted to approximately $490 \cdot 10^6$ m^3/year.

The general water quality pattern shows a decreasing salt content with depth. Generally, the aquifers contain less than 2000 ppm total dissolved solids and are carbonate rather than chloride enriched.

As no significant marine transgression took place in the basin until the Lower Cretaceous, the Jurassic aquifers may have been shielded from sea water intrusion by the high head of the fresh water already flowing into them. The Cretaceous aquifers on the other hand were fully exposed to saline water. This fact may in part account for the vertical variation in salinity.

Another large basin with similar hydrogeological characteristics is the Sahara Basin which extends along the northern margins of the African and Arabo-Nubian shields.

The aquifers of this basin are composed of continental sandstones of Lower Cretaceous age, underlain and overlain by marine layers of marls and limestones. In the Central Sahara the aquiferous sandstones are termed "Intercalcaire Continental", while in the Western Desert of Egypt they are called "Nubian Sandstone" (Knetsch).[5]

As the Sahara is bordered to the north by the Atlas Mountains and in the south by the Ahgar crystalline massif, the hydrological regime resembles that of the Artesian Basin of Australia. The isopiezometric maps show the flow directions to be from the northern margins of the Sahara, i.e. from the southern margins of the Atlas range toward their central portion. The outflow is by seepage to the surface, by evaporation or by outflow into other aquifers and thence into desert mudflats. Along the margins, the aquifers are phreatic while in the central parts of the desert they are confined and artesian. The volume of water estimated to be stored under the Sahara is claimed to be $(4 \pm 1)\cdot 10^{12}$ m^3 according to Cornet and Rognon.[6]

The Western Desert of Egypt is bordered to the north by the Mediterranean Sea, and flow occurs mainly from south to north. The discharge areas are those where gentle folds bring the sandstones near the surface. In these areas, the water table is artesian and the water finds its way to the surface through joints and fault lines where the oases of the Western Desert are found (Knetsch).[5]

The recharge of this basin occurred mainly during the Pleistocene period when more humid conditions prevailed in this part of the world. Age determination tests using ^{14}C showed the age of the water to be several tens of thousands of years old (Munich and Vogel).[7]

The water in the southern parts of the basin is relatively fresh and it becomes more and more mineralized as one progresses further into the basin. North of the oases, the water is highly mineralized probably due to the fact that this part of the basin was not flushed by fresh water, of the sea water which filled it during the Cretaceous and Eocene transgressions.

The continuation of the Western Desert basin is found in the Sinai Peninsula and southern Israel (Issar, et. al).[8] In this region evidence exists for some $2\cdot 10^{11}$ m^3 of fresh to brackish water lying within the sandstone layers of Lower Cretaceous age.

One also finds aquifers containing reserves of fossil water in the thick alluvial deposits found along the foothills of mountain ranges. Such aquifers are found along the southern margins of the Alborz, Hindukush, and Himalayan Mountains as well as along the eastern margins of the Rocky Mountains and the eastern portion of the Andes.

The thickness of the alluvial deposits amounts to several hundred meters. Although the water found in them is replenished annually in the upper reaches of the alluvial fans, the distance between the recharge zone in the foothills and the regions of pumpage in the plains may exceed hundreds of kilometers, and thus the water under the plains may be regarded as fossil for all practical purposes.

Other vast fossil aquifers or non-replenishable storage may be found in limestone aquifers such as those of the Zagros Mountains of Iran.

In most of the regions described above the only alternative resources are deep-lying saline aquifers. As this water involves deep drilling, pumpage, and desalination, it is better to disregard them as alternatives.

III. CONCLUDING REMARKS

The problem of non-replenishable resources has to be solved as part of a systems management approach, namely by regarding the region as a production system and by defining the physical components within it. Thus the volume of the extractable fossil water can be introduced as one of the constraints together with others on the system.

By systems analysis methods the policies required for the long-term operation of the system can be achieved. The objective of the system should be incorporation of the one-time reserve into the stochastic system of surface or shallow ground water resources, or in other words, the optimization of the annual water supply to give the highest steady income. Thus, the objective function can be expressed by the net benefit of the proposed system.

A basic problem in the planning of such systems is the uncertainty factor, which is due to the randomness of the supply of surface water on the one hand, and the lack of hydraulic and chemical coefficients of the fossil aquifers on the other. In this case, methods of planning under uncertainty have to be introduced. This involves making the preliminary plan with the information available, and at the end the results of the unknown coefficient will be checked by parametric analysis.

In those cases where some hydrological data are available, the coefficients can be derived by a simple procedure of simulating the hydrological observed data with synthetic data, and by trial and error of the coefficients until an agreement between them is reached. This procedure can be applied only where a simple hydrological model is considered and is in the preliminary stages of planning.

To conclude, it is recommended that non-replenishable aquifers, especially in arid regions, be regarded as a factor in development, and that they be included in regional water development projects as a steady supply to compensate for the uncertainly of other water resources. This is on the condition that conventional procedures of systems planning are followed.

IV. REFERENCES

1. M. Evenari, L. Shanan and N.H. Tadmor, The Negev: The Challenge of a Desert (Harvard University Press, Cambridge, Mass., 1971).

2. Study of Water in the Chad Basin (UNESCO 1969)

3. M.C. Hind and R.J. Halby, J. Geol. Soc. Aust., 1, 16, 481-497 (1969).

4. Water Resources Development in Australia, New Zealand and Western Samoa,(United Nations, 1968).

5. G. Knetsch, Geolog Rundschau, 52, 640-650 (1962).

6. A. Cornet and Ph. Rognon, B.R.G.M. Chronique d'Hydrogeologie, No. 11, 82-96 (1967).

7. K.O. Munich and J.C. Vogel, Geolog. Rundschau, 52, 611-624 (1962)

8. A. Issar, A. Bein and A. Michaeli, J. of Hydrology, 17, 353-374 (1972).

SOLAR ENERGY FOR DESERT SETTLEMENTS

DAVID FAIMAN

I. HISTORICAL INTRODUCTION

If the harnessing of solar energy did not actually begin in the desert its use in that milieu certainly dates back to the earliest traces of recorded history. The biblical book of Joshua refers to Misrephoth-maim, a place where, according to tradition, the sun's heat was used to extract salt from sea water. Moreover "Salt of Sodom" is an expression that was current in the Palestine of Roman times.[1] That same Sodom, located on the shore of an inland salt lake, is today the site of Israel's Dead Sea Works Corporation, where solar energy is the driving force for more than 100 sq. km of evaporation pans.[2]

The term "driving force" as used here is no mere platitude, for in the course of a year the amount of solar energy that goes into the evaporation process in these pans is equivalent to some 25 million barrels of oil.

There is another way of looking at all these barrels of sunshine, for in the course of a year they evaporate 350 million cubic metres of water from the 100 sq. km. of Dead Sea pans. Serious attempts at using solar energy for obtaining pure water had of course to await an understanding of some elementary principles of physics, but their roots also lie dimly buried in ancient history. It is claimed[3] that the Nabateans of two millenia ago built net-works of rock piles - the so-called "Teleilat el-anab" that dot the Negev - in order to enhance the formation of dew. Actual solar stills on the other hand, have been identified[4] as far back in history as a system of 4700 sq. m. constructed in 1874 in Chile. This particular still produced[4] about five litres of fresh water per day for each square metre of surface area on clear sunny days. This represents about the best performance one can expect from a still whose sole function is to use solar energy to evaporate water and then waste the energy during the condensation stage. Nowadays, research effort is being concentrated into the so-called multi-effect type of still. This device harnesses part of the latent heat released upon condensation and uses it to pre-heat the incoming saline water. To date, daily yields of up to 19 litres per sq. metre[5] have been obtained this way and the method clearly shows much promise for future desert settlements.

Production costs will however, have to come down and long term reliability of operation be demonstrated.

Yet another use of solar energy whose origins are lost, but which conceivably originated in the desert, is fruit drying, or more generally crop-drying. Today the Bedouin of Sinai store sun-dried dates for out-of-season use and it is likely that their Nabatean precursors in the region did likewise. Here too modern day research is trying to render solar crop drying cost-effective for contemporary farm outputs. Experiments range from low-cost short-lived plastic sheeting[6] that traps solar radiation and heats an air stream to the high-cost long-lived collectors[7] that we shall consider in Section III below. Once again the main obstacle to be overcome is the cost barrier.

This brief historical survey illustrates the technological simplicity of solar energy utilization and its suitability for remote desert regions. It also teaches us an important lesson, namely, the fact that solar technology did not begin to make significant advances until the third quarter of the twentieth century was a simple consequence of the comparative cheapness of other, more convenient, forms of fuel. That solar fuel has once again started to look economically promising presents us with two problems. First how to retrofit existing desert settlements in order to save on fossil fuel, and secondly, how to design future settlements to make maximum use of this commodity. These two problems are dealt with in the following sections.

II. SOLAR RETROFIT

The Negev desert contains several settlements that are two decades or more in age. They hail from an era of cheap imported oil, with the result that housing and plumbing have little if any in the way of thermal insulation and domestic water is often heated by electricity. Three aspects of this sorry legacy can be treated via existing technology with varying degrees of economic attractiveness. They are: single-family domestic hot water, community central hot water supply and winter space heating.

a. Single Family Domestic Hot Water

The most common kind of hot water supply for individual families consists of a 120 litre water tank containing an electric immersion heater. Figure 1 shows the monthly average daily electricity consumption of one family with two children.

The figures are based on electricity meter readings taken over a five-year period. At current electricity prices of $0.06 per kWh the annual electricity bill for this family's hot water comes to $118.

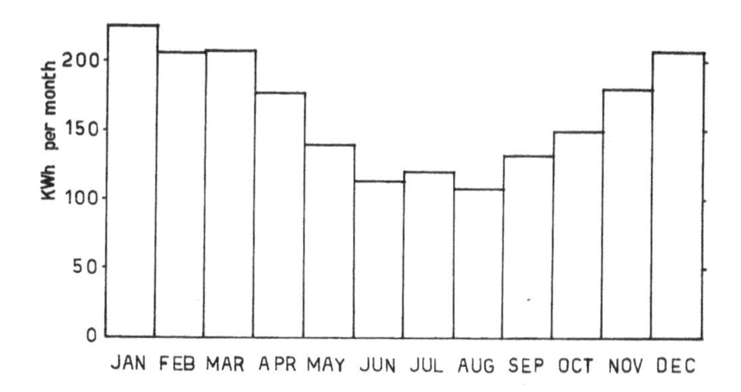

FIGURE 1, Monthly average daily electricity consumption for water heating for an actual two-child family in a Negev settlement.

Now a simple retrofit can be (and was) made.[8] For about $400 one can purchase a further 120-litre water tank and a solar collector panel of between 1.5 and 2 sq. m. effective area (Figure 2). The idea is to use the solar collector to heat water in the new tank and then to pass this pre-heated water down into the existing electrically heated tank in the household. The electric heating element in the latter is of course thermostatically controlled so that on sunny summer days no further heating will be necessary. The arrangement is shown schematically in Figure 3.

FIGURE 2, One-family domestic hot water solar retrofit.(Photo: A. Bar-Lev)

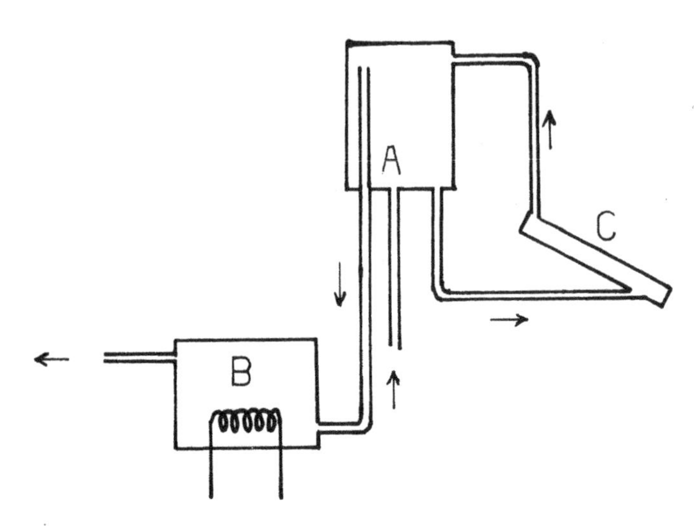

FIGURE 3, Solar retrofit of single family with existing electric immersion heater tank B. Cold water supply enters new tank at A, is heated by the solar collector C and then passed down to tank B where it is further heated (if necessary) by electricity.

By using standard analytical techniques[9] one can estimate the month-by-month electricity savings for such a system. These are shown in Table 1 for the specific family and hardware in question.

TABLE 1, Monthly electricity savings with a solar retrofit system.

Month	Jan	Feb	Mar	Apr	May	Jun	Jul	Aug	Sep	Oct	Nov	Dec
% electricity saved	44	48	56	68	83	97	96	100	89	79	62	42

By comparing Table 1 with Figure 1 one arrives at the conclusion that 67% of this family's domestic water heating bills will be saved; namely $79 per year. At this rate the system will have paid for itself in about five years. And notice that the family now enjoys twice as much hot water!

b. Community Hot Water

Many Negev settlements supply hot water to their inhabitants from a central oil-fired boiler. One particular settlement was found to be using 8000 litres per day (in winter) of water at 62°C and running up an annual fuel bill of $3,000. A detailed study was made[10] and as a result the solar retrofit shown in Figure 4 was installed at a cost of about $10,000. The expected monthly fuel savings are shown in Table 2.

FIGURE 4, Solar collector array at a Negev kibbutz. (Photo: A. Bar-Lev)

TABLE 2, Monthly fuel savings with a solar retrofit system at a Negev kibbutz.

Month	Jan	Feb	Mar	Apr	May	Jun	Jul	Aug	Sep	Oct	Nov	Dec
% fuel saved	46	49	51	60	86	100	100	100	100	90	60	40

The annual fuel savings are predicted to be 62% or $2,100 leading once again to a payback period of about five years.

c. Winter Space Heating

The last example we present for technically feasible solar retrofitting involves winter heating. Here the main problem to be overcome in most Negev buildings is the poor thermal insulation in the building envelope. The first stage of tackling the heating problem thus does not involve solar energy at all. It involves sealing cracks around windows and doors to prevent draughts, double glazing the windows by adding another pane of glass with a 2 cm air gap between it and the original window, and adding thermal insulation to the walls and roof. If this job is done well it could turn out that little further heating is necessary apart from the occasional use of a small electric heater on exceptionally cold days.

If the climate is such that substantial heating is required then the economics become extremely subtle. As far as solar energy goes such heating may be effected "actively" or "passively". The former method involves the use of solar collector panels, pumps, thermal storage (for after sun-down heating) and a heat distribution system (air ducts, radiator panels etc.). Two factors render "active" systems economically unattractive at present. The first is the large area of collector panels required: typically a substantial fraction of the floor area of the house. This drawback may be overcome in time by the emergence of cheap mass-produced all-plastic solar collectors.[8] The second problem is that this expensive heating equipment lies idle for most of the year causing its amortization period to be artificially extended. This drawback might be overcome one day if the system could be used for cooling purposes during the summer. Once again however, a lot of fairly sophisticated technology will have to come down in price by a substantial amount before "active" solar all-year space conditioning becomes economically attractive.

"Passive" heating is by far a more attractive proposition. It is discussed in greater detail in the chapter by Givoni. For our purposes it is sufficient to note that here the building is made to act as its own solar collector. Of the various methods discussed by Givoni one is particularly suitable for a solar retrofit to an existing building, namely the add-on greenhouse.

As an example we may cite a nursery school located in a certain Negev settlement. Such public buildings are particularly attractive candidates for solar heating owing to the fact that they are not used at night and hence no thermal storage is necessary.

On account of the design and construction of the nursery school in question, it was decided to add a greenhouse to the southeast facing wall in order to heat two classrooms. The glazing material consists of a double-walled poly-carbonate material containing a 7 mm air gap trapped within to provide thermal insulation. This is supported by a wooden framework designed in such a manner that it can be opened for ventilation purposes during the hot season. The interior of the enclosed area is painted in a dark colour to absorb radiation, and contains plants and water barrels to prevent over-heating. The actual amount of daily solar radiation falling on this greenhouse in January has been calculated[8] to be 360 kWh of which 30% may conservatively be expected to contribute towards heating the 150 sq. m. of classroom floor area. The construction cost will be about $3,000. For a four-month heating season and an arbitrarily taken five-year lifetime, this investment represents about $0.025 for each kWh of useful heat. Figure 5 shows the exterior of the two classrooms prior to the glazing over of their porches. The present heating system comprises a 10 kCal/hr kerosene stove in each classroom. If these stoves are retained as back-up, the solar retrofit is calculated to save 1620 litres of kerosene per heating season.

FIGURE 5, Southeast aspect of a Negev nursery school prior to "greenhouse" retrofit.
(Photo: M. Cones)

III. THE SOLAR FUTURE

Lest the reader suspect that we are about to embark now upon a speculative overview of what science might hold out for the years to come, let it be stressed from the beginning that the dividing line between this and the preceeding section is a hazy one. It is in fact mainly one of economics, for whereas all the systems discussed in Section II are of proven cost-effectiveness those to be discussed now will have to come down in price to a lesser or greater extent. They are however, all technologically feasible and have already been demonstrated. Some of them indeed are even cost-effective in certain situations. Moreover, with production techniques constantly leading to lower prices, it is likely that desert locations will be the first to benefit from the new generation of solar devices - and soon.

Accordingly, we shall now discuss - probably in chronological order - some potential applications of solar energy to desert settlements.

a. Optimizing the Back-Up

As if to emphasize our point on the haziness of the dividing line between this and the previous section, we start with a totally soluble - both economically and technologically - problem. The systems discussed above were solar retrofits in which solar collectors were added to existing systems in order to cut down on fossil fuel consumption. In each situation the emphasis of the research was on how to input solar energy to an already existing back-up system in as cost-effective as possible a manner.

For desert settlements of the future, namely all those at present under construction and the ones yet to be planned, the question must be put the other way around. For we know that solar heated water is economically justified. What we do not yet know is what back-up system is best "for a rainy day".

The answer will of course, vary from one location to the next and probably from epoch to epoch. For example, at present in desert locations, electricity is the most expensive source of energy. At face value therefore it would appear to be unwise to install large-scale electric boilers for water heating. If however, we are thinking of remote sunny locations, which the deserts are, an electric back-up might turn out to be extremely sensible. For on the one hand electricity probably presents the most maintenance-free form of heating in existence. On the other hand, as we shall discuss below, solar-generated electricity will soon be available.

Similarly, in certain locations, natural gas might yield the cheapest form of back-up. It is thus clear that only when design of solar system and back-up are undertaken together, a truly economic optimum can be arrived at. Looked at from another angle, the advent of cheap solar energy will cast a different perspective on the relative economics of conventional energy sources.

b. Process Heat

The most sophisticated solar heating systems we have hitherto discussed are based on so-called "flat plate solar collectors". These are essentially pre-Industrial Revolution technology with an optional twentieth century improvement[11] to their thermal properties. Their limiting feature is that above about 100°C they cease to convert useful amounts of solar energy to heat. In order to proceed to higher temperatures, two methods are now becoming commercial.[7]

One employs the use of large-area lenses or mirrors to concentrate the sun's rays on to a comparatively small area. The other utilizes high vacuum techniques in order to reduce heat losses and so enable the solar collector to operate at higher temperatures. There are even collectors which combine both of these principles.

At present these kinds of collector cost about $300 per sq. m. installed in typical U.S. industrial demonstration projects. These prices will however, come down , both in the United States and particularly in countries where labour costs are lower. Nevertheless, in spite of these high costs, it is instructive to look at two potential uses for this kind of technology in desert settlements of the future. Of particular interest in this respect are modules of evacuated tube collectors which, since they have no moving parts, hold out every prospect of being relatively maintenance free.

i. Algae Drying

The potential importance of an algae industry for desert settlements has been discussed in the chapter by Richmond. One limiting factor is the enormous energy consumption that is of needs associated with the drying process. Specifically, in order to obtain 1 kg of dry algae one must typically evaporate about 5 litres of water. This requires the expenditure of some 11 MJ of energy (approximately 3 kWh in electrical units). Unfortunately for most purposes the drying needs to take place at about 120°C so one cannot simply put the algae out in the sun to dry.

Turning however towards the new generation of high temperature solar collectors, one square meter of such technology is capable of contributing about 2,400 MJ or 670 kWh per six-month growing season. Thus, even at their currently expensive price of $300 per sq. m., over a claimed 15-year lifetime, the energy works out at $0.03 per kWh.

Algae drying is a process particularly suited to solar energy for two reasons. Firstly, the algae growth rate is correlated with the amount of solar radiation available. Hence, most energy is available for drying at a time when most algae need to be dried, and vice-versa. Secondly, the fact that the solar energy is to be used as it is collected and not stored for later use will in general lead to relatively high conversion efficiency. If solar energy has to be stored there are necessarily losses incurred in getting it into and out of storage and thermal losses from the store. But this is a technicality that need not concern us here.[12] The main point is that algae drying is one of several potential desert uses for this new generation of high temperature collectors in which their useful energy output can be expected to be relatively high.

ii. Steam Production

Our second example of a potential use for this kind of collector involves steam production in conjunction with a back-up system. The distinction between this and the previous example is that here we are talking about a situation in which steam or process heat is required at times when the sun might not happen to be available.

To be more specific, we may consider a certain Negev settlement for which a solar steam retrofit is currently under consideration. This settlement uses steam (generated by an oil-fired furnace) for its communal dining hall and laundry. Its annual fuel bill for this facility is at present about $7,800 for which the boiler puts out 350,000 kWh of energy. This works out at about $0.02 per kWh which is not too different from the cost estimate of solar heat we derived in the previous example. Needless to say, the economics of such schemes need to be carried out much more carefully, but with rising oil prices and falling collector costs, the time is clearly not too far off when solar steam production will be a substantial money saver for desert settlements.

c. Solar Electricity

A promising device to have emerged from the solar age is the photovoltaic cell (see Fig. 6). This coin sized object with two wires connected produces a continuous electric current when exposed to solar radiation. Indeed, an area of 1 sq. m. of such cells joined together is capable of producing about 100 watts of power when exposed to the noon-day sun on a clear summer day.

At present, such cells are extremely expensive to produce, but there is an intensive effort going on, with every prospect of success, to bring their price down to $50 per sq. m by 1985. Moreover, a recent study carried out by the American Physical Society[13] concludes that arrays of such cells will compete economically with other forms of energy if they can be produced for between $10 and $40 per sq. m.

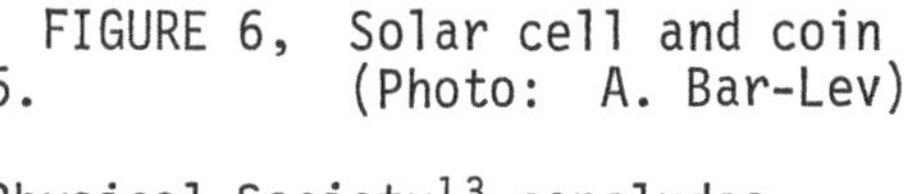

FIGURE 6, Solar cell and coin (Photo: A. Bar-Lev)

To see what this means in practice for our desert settler, let us return to the family whose domestic water heating needs were discussed in Section IIa above. Approximately half of this family's electricity consumption went into heating water (before the retrofit!). That is to say, their annual total electricity consumption was about 4000 kWh per year or $240 per year at present day electricity costs. (It is actually now 2/3 of this figure but let us ignore this correction so as not to complicate the present discussion). Now this quantity of electricity could be supplied by 20 sq. m. of photovoltaic cells which in 1985 should cost $1000.

This is not quite all the story because no allowance has been made for the fact that these cells cannot produce electricity at night. Thus, storage batteries will be necessary and at present these are expensive. We have also ignored the fact that photovoltaic cells produce direct current, whereas most domestic appliances require alternating current. Thus, a DC/AC converter will be needed. There is no doubt, however, that these and several other technical obstacles will be overcome leading to profound consequences for desert settlements.

In the first place, paradoxically, electricity will be "re-discovered." After more than a decade of trying to find methods to cut down on the use of this commodity, electricity will again be available in ample supply. Thus, for example, winter heating and summer cooling of the home will be possible using "old-fashioned" electric heaters and air conditioners. It is a sobering thought that regular air conditioners powered by solar-derived electricity might become cost effective before solar air conditioning technology is fully developed!

Also of major importance will be the fact that desert settlements will be possible without the expense of laying kilometers of power lines.

d. Saline Ponds

Another promising development in the direction of harnessing large quantities of solar energy for desert settlements is the non-convecting solar pond. This is an expanse of water which is artificially maintained in a state such that its salinity (and hence density) increases with depth. Solar radiation which penetrates to the bottom of the pond heats the nearby water which is prevented from rising and mixing by its greater density. In this manner, saline water at temperatures close to 100°C has been obtained.[14] The great attraction of this kind of device is two-fold. First, it is possibly the least expensive method of producing large-area solar collectors. Thus, even if the Carnot efficiency is relatively low the heat capacity is enormous and serious consideration can be given to these ponds as a source of power. Indeed, a fairly large installation has been reported[15] that will supply power to a hotel on the shore of the Dead Sea.

The second attractive feature of the saline pond as a solar energy collector also arises from its great heat capacity, namely that it acts as its own energy storage device. Night and day radiation differences do not influence the temperature at the bottom and seasonal changes only have a slight effect.

Thus, if technical difficulties are overcome, a saline pond of area 100 km^2 located in the Dead Sea could produce a continuous electricity output of 2000 megawatts, and the Dead Sea is large enough for several such ponds.

IV. CONCLUSIONS

In the previous sections, we have presented a survey of the manner in which solar energy offers considerable promise as a serious aid to the implementation of desert settlement. We must emphasize that this is not intended to be a review of all solar energy applications. In particular, we have omitted any serious discussion of two major areas of solar research. The first is of course, all the "classical" uses of solar energy briefly alluded to in the introduction and a few others, including wind power (an indirect form of solar energy). These techniques have been researched for many years, and to do them justice would require a review far longer than that which is our purpose at present.

The second major area of research effort that we have ignored for our purposes concerns the large-scale production of power. It is true that with the depletion of fossil fuel reserves, the deserts of the world might one day provide solar power stations for the more populated areas of the globe, but this is an issue for the distant future and a controversial one at that. The reality of such a possibility entails such considerations as whether solar energy beamed down from extra-terrestrial satellites[16] to small earth-bound receiving stations might not be more effective than large-area desert collectors[17] and a host of technical considerations.[18] Furthermore, there are the timetables for the realization of other non-fossil forms of power such as nuclear fusion[19] which have to be considered.

Instead, we have concentrated on the problem of settling man in the desert and providing him with all the comforts of the more temperate regions with the aid of solar energy.

FIGURE 7, Central receiving towers illuminated by thousands of mirrors might one day supply electric power to the world's deserts.
(Photo: courtesy of EPRI.)

We have given specific examples (in Section II) of the methods by which present-day technology is already cost effective in diminishing the fossil fuel consumption of existing desert settlements. Of course, many of these examples are not of exclusive applicability to desert conditions; and fortunately so, for today's desert settlers are in no way the major consumers of energy on this planet. Rather, solar retrofitting of existing settlements should be viewed by policy-makers as an easy method of obtaining experience with solar energy, and by the settlers themselves as a method of cutting down on the cost of living.

The second aspect of solar energy to which we have addressed attention (in Section III) involves the planning of future desert settlements. Here we have invoked technology which is only of marginal cost effectiveness at the present time. But it has taught us a very important lesson, namely that of the host of obstacles that must be overcome if the deserts of the world are to become seriously populated, power production is not one of them.

That is not to imply of course, that there is no energy crisis. Of course there is, and any population increase will exacerbate it if no alternative to fossil fuel is ultimately provided. Rather, the point to note here is that deserts by their very nature lend themselves to rendering their inhabitants energy independent. Thus, although the recent APS survey concludes[13] that photovoltaic cells can only supply a maximum of a few percent of the world's (actually America's) power requirements by the end of this century, this will not be true for desert settlements. Indeed, we have indicated that solar-generated electricity should be an important ingredient of a desert economy of the future.

It thus appears that our discussion has come full circle from flat-plate to flat plate. We started out from the tried and proven water heater of yesterday and ended with the modules of evacuated tubes and photovoltaic cells of tomorrow. In each case cheap power with no moving parts: a bright prospect.

V. REFERENCES

1. Talmud Babli, Kerithoth 6a.

2. E. Orni and E. Efrat, Geography of Israel (Israel Universities Press, Jerusalem, 1976).

3. N. Glueck, Rivers in the Desert (Grove Press, New York, 1960).

4. A.B. Meinel and M.P. Meinel, Applied Solar Energy (Addison-Wesley, Reading, Massachusetts, 1977).

5. C.N. Hodges, T.L. Thompson, J.E. Groh and D.H. Frieling (1966) (quoted with some details in ref. 4).

6. H.R. Bolin, A.E. Stafford and C.C. Huxsoll, Solar Energy 20, 289 (1978).

7. Putting the Sun to Work in Industry (S.E.R.I., Golden, Colo. 1979)

8. D. Faiman, unpublished.

9. W.A. Beckman, S.A. Klein and J.A. Duffie, Solar Heating Design by the f-chart Method (Wiley, New York, 1977).

10. D. Faiman, J.M. Gordon and D. Govaer, Israel J. of Technology 17; 19 (1979)

11. H. Tabor, Bull. Res. Council of Israel 5A, No. 2, 119 (1956).

12. D. Faiman, Solar Energy 19, 743 (1977)

13. Principal Conclusions of the APS Study Group on Solar Photovoltaic Energy Conversion (American Physical Soc. New York, 1979).

14. H. Tabor, Desalination 17, 289 (1975).

15. B. Doron, Invited paper at the International Conference on the Application of Solar Energy (Haifa, 1978).

16. P.E. Glaser, J. of Spacecraft and Rockets 13, 573 (1976).

17. A.F. Hildebrant and L.L. Vant-Hull, Science 197, 1139 (1977).

18. D. Faiman, Solar Energy 22, 397 (1979)

19. P.L. Kapitza, Rev. Mod. Physics 51, 417 (1979).

IS DESERT SETTLEMENT ECONOMICALLY VIABLE? THEORY VS. REALITY

URI REGEV

I. INTRODUCTION

In his paper on the philosophy of water management Thomas Maddock[1] put forward the hypothesis that any development of a desert which is not based on obvious, available, natural resources such as oil or minerals is doomed to failure, either by eventual starvation or by continuous subsidization by the outside world. Is it possible to settle the desert and make it bloom on a sound economic basis, or is continuous outside support unavoidable? I believe this is our real challenge.

Resource allocation in arid zones does not involve principles different from the general economic theory. It is rather an application of the general economic principles to the specific conditions of a desert. These special conditions usually include, among other things, water scarcity, great distances, high transportation costs, plentiful land and solar energy and sometimes mineral resources. As a result of these conditions, desert areas in underdeveloped countries are largely unpopulated: their meagre populations being usually low in income and standard of living. This however, is not always the case in developed countries where one can find cities and industries in the middle of deserts. The difference between these two cases is the key to meeting the challenge of desert development. The economic basis for Maddock's statement will be clearly seen when the desert's economy is based on its natural resources of little and highly variable water renewal and poor soil, while using conventional agricultural technologies. The small amount of renewable water leads to agricultural technologies which are land using and water saving, and thus the carrying capacity of the land is low. Economically this would imply relatively cheap land, high water prices and sparse agricultural population per unit area. Urban centres are unlikely to develop because of the adverse climatic conditions and large distances from other population centres. Special local conditions, such as specific microclimate control points for commerce roads, and special mineral sources gave rise to some thinly dispersed urban centres. Such an economy could support only subsistence levels of income that would leave sufficient reserves for the climatic variability and specifically for drought years. Any attempt to alleviate the desert population under these circumstances and technologies, is indeed doomed to failure. The major reason is that of precipitation variability, which in general is negatively correlated with its quantity. Agricultural production, therefore, involves higher risks which are accounted for by the large land requirement per capita. Increasing population without outside support increases the probability of destruction in the 'game against nature' by reducing the reserves and by violating the ecological equilibrium necessary for food production under these conditions.

The prospects of desert settlement in developed countries are, however, different. Highly concentrated urban population adjacent to a desert area leads to new economic incentives of desert settlement:

(1) Increasing land prices around population centres emphasize the relatively low price of the desert land which makes it more attractive for land using industries.
(2) Developed infrastructure lowers the transportation costs which otherwise hinder development of peripheral regions.
(3) Industrial technological changes enhance investments in capital intensive and water saving production - alternatives which can capitalize on the relative advantages of desert areas, e.g. by using solar energy.
(4) New agricultural techniques, such as closed system agriculture could be developed, which is both water saving and capable of using the abundant solar energy.
(5) Pollution and environmental costs produce additional incentive for migration to peripheral regions, both for individuals and firms which have otherwise to pay increasing pollution costs.
(6) So-called 'non-economic' objectives for desert settlement of developed countries, such as defence, population dispersion, etc.

II. THEORETICAL CONSIDERATIONS

The basic economic principle relevant to our problem is that of marginal cost pricing. In its simplest form it states that economic efficiency is attained when resources are priced at their marginal costs and at this price they are available to everyone who is willing to pay. Since water scarcity is the major factor which determines the existence of most desert areas*, it is appropriate to demonstrate the application of this principle to water price. The marginal cost of water, as any other scarce resource is composed of (a) the cost of bringing the last unit to the location where it is needed, and (b) scarcity of alternative cost. As for the first part, if water is transported from one area to another its cost differs exactly by the transportation cost. Thus, once a desert area receives water from outside sources, its economic price should be higher, and this difference is greater, the greater the distance from the water source. Secondly, if water is obtained from local sources only, its scarcity costs are bound to be higher in a desert area.** In addition to that, both living and food production in a desert require higher water use than in a humid area. Therefore, given free competition of regional production technologies, it becomes clear that the same (agricultural) product can be produced with lower costs in non-desert regions. This could be considered as the main reason why desert areas are sparsely populated in the first place. Thus, if water is to be priced at the marginal cost, conventional production technologies would not be able to pay this price in a desert area. This implies that, disregarding other factors for the moment, development of a desert area could not be economical with high water transportation costs and water requirements that are higher than in non-desert areas.

Another economic disadvantage of a desert area is its distance from population centres, and distances within the desert itself. Transportation costs are another major detriment to the location of industries and urban centres in desert areas, and thus constitute an additional hindrance to the development of these areas. However, the

---

* There are, of course, deserts of ice surrounding the north and south poles, where low temperature rather than water, is the main problem. These are excluded from our discussion.

** The difference between water prices in two adjacent areas can not be larger than transportation costs, even if transportation does not take place.

infrastructure of roads could change this situation considerably by reducing these costs. In most cases, both in the past as well as the present, roads going through deserts were built for reasons other than desert development - commerce, defence, etc., and along these roads, desert settlement becomes both economically and socially easier, though still more costly than near population centres.

Though water scarcity and long distances discourage desert settlement, there are additional factors which, at times, could play a major role. Among these are adverse climatic conditions, isolation, risk and uncertainty.

The first two are outside the range of our discussion. Uncertainty and variability of natural conditions are far greater in the desert than in more humid regions. It has been shown[2] that for risk averse decision makers, uncertainty leads to lower output as compared to deterministic conditions. Moreover, risk aversion also implies that people in general would prefer the more stable and humid regions rather than a desert.

The above economic considerations are not the only ones which deter people from living in the desert. However, the social, psychological and other considerations are beyond the scope of this paper, and thus are not discussed here.

However grim, this is only one side of the coin. There are several economic advantages of a desert area. The first and most well-known is the availability of minerals, oil and other natural resources. If and when these are found in quantities, it is obviously a basis for development - Saudi Arabia is but one example. However, there are some additional economic incentives for migration to an arid or semi-arid region. Land abundance and lower economic alternative use imply a low (and possibly zero) price of land in a desert area. This makes it desirable for technologies which use land intensively but do not greatly suffer from the above mentioned disadvantages. Likely candidates for that are recreation activities. Modern times have created an increasing demand for recreation and the unpopulated desert areas can serve as a source of supply for that need. But more important effects of our time on the potential for desert settlement, results from (a) population explosion, (b) environmental pollution in population centres, and (c) technological innovations which enable us to use the climatic advantages of the desert.

The first point does not need much elaboration. Congestion is a major incentive for migration to peripheral regions, and thus an economic basis is created in regions closer to the desert. This diminishes one major disadvantage - distance and transportation costs and enhances the potential for development of the remoter regions. Population density has a second effect - it increases pollution damage by certain industries. For example, a polluting factory, which was located away from the centre, gradually becomes surrounded by an increasing number of people who suffer from it. There has been an extended controversy over the issues of pollution and property rights and damage liabilities (e.g., Coase[3]; Baumol and Oates[4]). But irrespective of this issue, it is clear that larger population results in higher pollution costs to society which seeks even remoter locations for such industries. A desert is a likely candidate for locating polluting industries, as it is characterized by thin population. This however, raises other problems, as pollution decay may be slower in arid zones. As an example, it could serve as a potential location for atomic energy reactors if it is sufficiently close to a sea from which salt water could be transferred by a canal.

The largest potential for development of arid zones depends upon technological innovations. By revealed preference, it could be argued that all the advantages and disadvantages sum up to a negative economic total. If this were not true, deserts would have been developed by now, without government support, but through the economic mechanism of profit incentives. Especially in under-developed countries, neither population explosion nor high pollution has automatically resulted in migration to desert areas, and Egypt is but one example. Technological progress and the energy crisis are capable of changing that picture. Increasing energy costs underline the importance of a desert as a source of solar energy. This technology is advantageous in the desert, for its two main abundant resources - land and solar radiation. This could imply both direct energy production and other production processes which could use the solar energy in the production process, such as in greenhouses for closed system agriculture, and for water desalination. Such progress, however, is expected to affect desert settlement in the long run when these technologies are more advanced, less expensive and when other costs (energy, pollution, etc.) become still higher than they are at present.

III. APPLICATION: DEVELOPMENT IN THE NEGEV

The development of the Negev has been among the goals of all governements in Israel, and its justification was based on (a) defence (b) population dispersion and (c) economic potential. In the short history of the State of Israel, a rapid development has characterized the Negev until 1965. By that time the natural obvious resources of the Dead Sea, phosphates and copper were developed and its population reached 150,000 (from 45,000 in 1954). Since then the development process has slowed down considerably. In "The Land of the Negev" (Shmueli and Grados, Hebrew)[5] a detailed description of the various aspects of the Negev, including some economic analyses can be found. Future plans and prospects are described in a "Council's Report - a Master Plan for Development of the Negev,"[6] and in "A Plan for Development of the Negev."[7] It is not our intention to analyze in detail these plans, but to shed some light and gain some insight as to the economic implications of future development in the Negev.

The first necessary condition for any development is water development. The potential additional water for the Negev could be obtained by additional water transfers from the north, brackish water from Tel Aviv metropolitan area, water desalination, and fossil water existing in large quantities in the Negev. The latter are (a) non-renewable and (b) need to be pumped from depths which are, in most parts, still beyond current technologies. Water transfers from the north become more and more expensive as water demands exceed the available renewable yearly amounts (around 1500 million cubic meters) and thus the alternative costs of water are very high. Desalination is probably the most expensive solution, so brackish water seems to be the first in line. Any of the above alternatives implies costly water far above the ability of conventional agriculture to pay*. Thus additional agricultural development in the Negev could not be economically viable unless based on new technologies or new crops which could capitalize on the advantages of the area. This could be said also about some of the existing agricultural activities in the Negev, but any attempt to change the situation needs to supply alternative employment and answer several social and political questions that would arise from it.

---

* The Arava region is an exception, since natural conditions are tantamount to glass house conditions elsewhere.

The peace agreement with Egypt has considerably changed the prospects and plans for development in the Negev. The region's main advantage of huge land reserves has vanished with the need to secure large areas for defence purposes, for air bases, camps and training areas. This has brought to the surface and sharpened conflicts between potential uses of the land - urban and industrial, recreation and parks, and defence.

The economic considerations constitute only one aspect in the multi-goal objective in the development of the Negev. Social, security, demographic and geopolitical aspects play an important part in the development decisions, and it is not an easy task to order them in their importance. One way to deal with the problem is to set some of the goals as constraints, e.g. population should be no less than a given number, or define a minimum quantity of land that would be allocated for defence. Then an economic maximization process should take place, subject to these constraints. This procedure is analogous to assigning weight to the conflicting goals and combining them in a single objective function.

The conflicts between defence and recreational uses amount mainly to vast land areas which could be allocated to parks or military uses. Both these uses are difficult to evaluate and the economic estimates of potential gains from recreation do not seem convincing enough in view of the pressing defence requirements for land. The allocation decisions are practically made on other grounds and considerations, but economic evaluation can still contribute to the process by adding its dimension to the problem. More specifically, identification of preferences for land areas for the two sides of the conflict could greatly improve the allocation by assigning lands with higher preference for parks, a flexible military use that would harm those least and leave the possibility of reversing the allocation.

The possibility of using new technologies based on natural advantages of the region is great indeed, as we have in addition to the arid conditions, also the advantages of the Dead Sea minerals with their potential, as well as the fact that the Dead Sea exists at 492 meters below sea level. The salt water carrier from the Mediterranean to the Dead Sea is one example of using the potential of these conditions. Once developed it can result in a major change in the prospects of development in the Northern Negev as well as the Dead Sea. Among others, it could make inland atomic energy reactors possible, a source for obtaining energy from solar pools and hydroelectric energy, tourism, industry and many more. Though it is not likely to start for 3-5 years, the allocation should be made now, as its economic potentials are extremely high, and its potential contribution to development of the Negev may be even higher through its indirect effects on industrial and agricultural activities that could use it to their advantage.

The situation is somewhat different when the transfer of industries from the centre to the Negev is considered. The distance which is the distinct disadvantage of the region is directly paid from the profits of the industry. The costs of being in the centre are however, not always paid by a factory. Specifically, the costs of land are not always paid, and the costs of pollution are hardly ever paid. Consequently, the economic incentives for moving away from the centre are hindered by the institutional framework and will not lead competitive firms to move. Government intervention is specially required in the case of pollution which constitutes an external effect, that under most circumstances cannot have a market price structure. This discussion implies that government support, required for transfer of certain industries, does not show its need for continuous subsidy, but is a way to amend the market failure to handle pollution of the environment. Caution should however, be taken in transferring polluting industries to the Negev, as desert conditions could magnify the problem of pollution decay.

Assessment of the potential economic contribution of development of the Negev cannot be complete without noting the possibility of mutual advantages of civil and military uses to each other. Infrastructure for one use is enjoyed by all others, and road construction for military uses is certain to diminish the disadvantage of costly transportation for industry and other economic activities in the region. On the other hand, development of services and certain light industries in the Negev will be enjoyed by the defence forces in reducing their costs. The contribution of the defence forces to the population and development of the Negev will however, necessitate a clear policy of government support in housing and supply of other services which will enhance the quality of life in the area. Without such a policy, these effects would be only marginal.

In sum, the economic success of developing a desert area, which is not blessed with obvious natural resources, depends upon:

(1) The possibility of water development at a reasonable price.

(2) Technological innovations which would enable the use of solar radiation or other arid zone advantages.
(3) Fairly developed neighbouring country which can support the infrastructure required as an initial investment.

These are the basic minimal conditions which will guarantee that deserts and arid zones could economically support income and living standards compatible with the rest of the country. The Negev region is not an exception, the stream of defence forces into it (following the Peace Agreement with Egypt) will neither block development nor initiate it automatically. Government policy could supply the basis for development, then its economic viability would depend on our ability to use its advantages - salt water carrier, winter crops in the Arava, further development of the Dead Sea for minerals and tourism, a parks system and the many potential developments of solar energy. All these enhance the belief that the Negev region will indeed bloom, contribute to the economy rather than be an economic burden, and attract investors, workers, tourists and leisure seekers with its clean environment.

IV. REFERENCES

1. T. Maddock, In: Arid Zone Development, Potentialities and Problems, Ed. Y. Mundlak and S.F. Singer (Baltimore Publishing Co., Cambridge, Mass., 1977).

2. H.E. Leland, American Economic Review, 62, 278 (1972).

3. R.H. Coase, J. of Law and Economics, 3, 1 (1960).

4. W.J. Baumol and W.E. Oates, The Theory of Environmental Policy (Englewood Cliffs, New Jersey 1975).

5. A. Shmueli and Y. Gradus, eds, The Land of the Negev (Ministry of Defense, Tel-Aviv, Hebrew, 1979).

6. S. Shachar, ed., Council's Report - A Master Plan for Development of the Negev (Ministry of Interior, Beer Sheva, Hebrew, 1977).

7. R. Lerman and A. Vachman, A Plan for Development of the Negev, First Interim Report,(Tahal, Tel-Aviv, Hebrew, 1979).

ECOLOGY AS A TOOL FOR DESERT MANAGEMENT

MOSHE SHACHAK

I. INTRODUCTION

Ecological studies of the desert are becoming more numerous and important as the population expands and the need for additional land increases.

In order to understand the relationship between settling the desert and desert ecology, it is necessary to define human settling of the desert in ecological terms. In this paper we will interpret the settling of the desert as dealing with the processes of introducing man-made systems into geographical regions that had natural balanced arid ecosystems.

Whenever man actively converts a part of a natural ecosystem into a man-made system, he must be concerned with the interactions that develop between the cultural and natural systems.[1] In the long run, these interactions determine the kind of relationship between man and his environment. The object of desert ecology is to gain insight into the structure, function and dynamics of the desert so that we can develop human systems that are as productive as possible and keep natural systems as diverse as possible. Exploiting the full potential of desert ecological research can help us approach this goal.

In this chapter we will attempt to show the significance of ecological knowledge in settling the desert. Our paper has two major objectives. First, to present three aspects of ecological studies that have been carried out in the Negev Desert; species richness, life history strategies, and ecological flow chains. These aspects of desert ecology are useful in dealing with various kinds of management problems including pest control, nature reserve management and understanding limitations in the long term efficient development of the area. The second objective emphasized is to identify some important ecological problems that require further investigation in the future.

II. SPECIES RICHNESS

Any desert settling affects the natural community's structure and composition. A desert community consists of many species with different population fluctuations and interactions with each other. Thus, the first ecological question should be concerning which species of plants and animals live in a particular desert community.

The aims of species richness studies are of interest for scientific purposes, nature protection, agricultural development and possible effects on inhabitants of the area. Man must understand the functioning of the system on the community level, protect species from extinction and know how species react to manipulations of the ecosystem.

The Negev Desert is diverse and rich in species of plants and animals. To illustrate species richness, the number of plant and animal species in the least diverse of our study sites - Sede Zin, are shown in Figure 1.

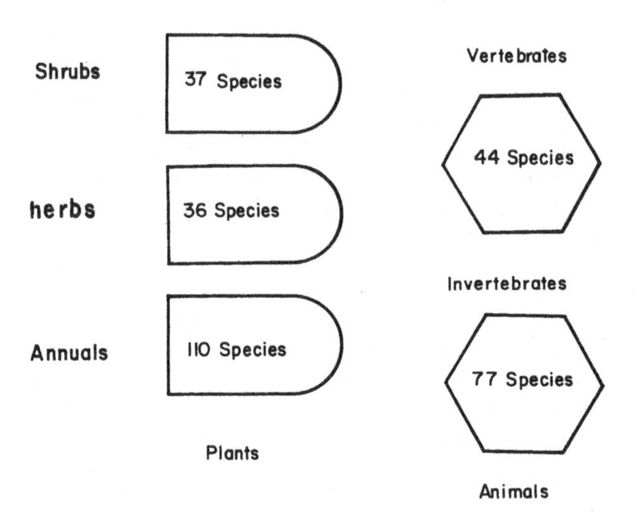

FIGURE 1, Species richness in the least diverse of our study sites - Sede Zin.

On a loessial plain in the Negev, over an area of 15 km^2 there are more than 300 species of plants and animals. Answering questions such as which of them is of value to man, which may become extinct, and which may become pests, comprise one aspect of interest for the desert ecologist that has a direct effect on the quality of life for potential settlers in the desert.

III. LIFE HISTORY STRATEGY

The consequences of desert settling are that man is constantly changing the structure and organization of the biotic community. What are the responses of the different populations to these changes? If a species becomes abundant it may have great economic impact as a pest. If a species decreases in number and becomes extinct it may have an impact on the stability of the natural ecosystem.[2] Knowledge of the life history strategy of desert species may help in the prediction of native population responses to man-made changes.

The key life history traits are brood size, size of young, the age, the age distribution of reproductive effort, the interaction of reproductive effort with adult mortality.[3] The general theoretical problem in the ecology of settling the desert is to predict the consequences of man's manipulation on the above traits. It is a very complex problem because there isn't one universal life history strategy for survival in the desert ecosystem. In dealing with this problem, it is important to keep in mind that a diversity of life history strategies have evolved in the desert. Therefore, under conditions created by man - reactions of populations with different life history strategies may be completely different. Some populations may increase and others decrease in numbers. One can illustrate the diversity of life history strategies that have evolved in the Negev desert by examining two species that coexist in the same ecosystem. These organisms are the desert snail *Sphincterochila zonata* and the isopod *Hemilepistus reaumuri*. As we shall see, their life history strategies are totally different from one another.[4,5,6,7]

a. The Desert Snail

Like most pulonate snails, *S. zonata* spend most of their life in dormancy and await favourable conditions before becoming active. The snails are active in the Negev desert only for a few days after a rainfall. While active, snails may feed or carry on other life functions such as locomotion, mating, egg laying and burrowing under bushes. During inactive periods the snails remain under shrubs either on the soil surface or buried to a depth of 4-5 cm. The desert snail's strategy is to become active for about 20-30 days and be dormant the rest of the year. Therefore, the snail is a desert animal which reacts directly to the two driving forces of the desert ecosystem, high radiation and low discontinuous stochastic water input. A model summarizing the strategies of *S. zonata* in coping with the high radiation load, lack of water and food, and uncertaintly of environmental conditions is shown in Figure 2.

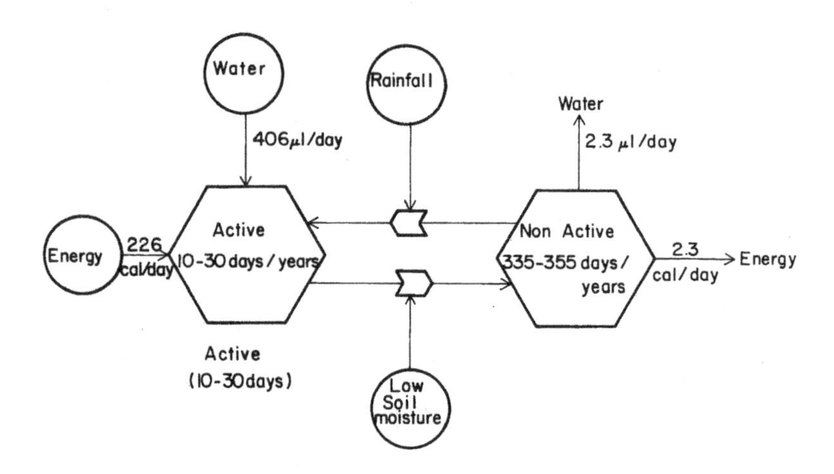

FIGURE 2, Adaptive strategies of the desert snail *S. zonata* - energy and water flows.

In order to cope with the problem of high radiation load, an interplay between behavioral and physiological adaptations has been developed. Toward the end of each period of activity, the whole snail population moves into the shade of the bushes and becomes dormant. During this stage, energy and water output are very low, 1.9 cal/snail/day and 2.3 μl/snail/day respectively. In the short active state, the snail is able to consume extremely large amounts of food (226cal/day) and water (406 μl/day). Comparison between water and energy flow in the active and inactive state shows that the snail can consume in one day of activity 800 times as much water and 300 times as much energy as it loses in a day of summer aestivation. Another way in which snails adapt to the uncertainty of the environment is by coupling reproductive activities with the rainfall pattern. As soon as activity is triggered by the first rainfall after the summer, a large proportion of the population mates, often before feeding begins. However, egg laying is more complicated and occurs only if the following conditions are fulfilled: occurrence of a rainfall at least one month after mating, ambient temperature higher than 4°C, and high soil moisture. The simultaneous occurrence of the above three conditions each year is of low probability. However, because of the long life span of snails (15 years) survival of the population is not at risk.

b. The Desert Isopod

The life span of the desert isopod *Hemilepistus reaumuri*, is one year. It lives in monogamous pairs with their offspring in completely closed family communities. The family lives in a subterranean burrow with only one entrance. Five distinct phases are recognizable in the life cycle of *H. reaumuri*; pair formation, gestation, hatching, growth and winter dormancy. Pair formation occurs from February until March. At this time the 9 month old isopods vacate the burrows in which they hatched. The females select sites for burrowing and begin to excavate new burrows. Pairing takes place after acceptance of a male by the female. During April, most of the females are gravid. In May, the young isopods hatch and are kept alive by parental care. The parents go foraging and carry food back into the burrows for the hatchlings. The growth phase is the longest in the life cycle of the isopods, from May to November. In this period a 30-fold increase in body weight was recorded. From November to February, no feeding or above-ground activity was observed. The ambient temperature is lowest at this time of the year and the isopods remain underground until pair formation.

Because of the evolution of the family community unit, the isopods have developed a life history strategy which is less directly dependent on radiation and rainfall. The family mode of life increases the probability of the individual's survival by parental care and co-operation between siblings. Parental investment which affects the survival of the offspring includes site selection for burrowing, protection of brood, and parental care during the early stages after hatching. Co-operation between the family members takes place mainly in two forms, expansion of burrow and guarding. A burrow can be dug only with the co-operation of at least 34 isopods. By utilizing the micro-climate of the burrow and the complementary above-ground activity pattern, the isopods are able to regulate their heat balance. The burrow micro-environment provides the isopods with comparatively high soil moisture. This increases air-water pressure and therefore, reduces isopod water loss. The regulation of radiation load and water balance can be achieved only by a high investment of energy by the family members in the construction of the burrow. The probability that the environment will provide the individual with its energy requirements is high. Thus, the utilization of high energy for digging on the family level is the solution for the individual water requirements.

The aim of nature protection in the desert is to prevent the extinction of such species as the snails and the isopods which have adapted themselves to a very hostile environment using completely different strategies. It is also found that the two species are actively involved in the soil turnover processes of the Negev desert. Mismanagement of just these two species may cause instability in the whole ecosystem. Figure 3 summarizes the relationship between species diversity, life history strategy, and consequences of mismanagement. One of the roles of desert ecology research is to investigate the diversity of life history strategies in order to predict the consequences of man's manipulations and to suggest how to avoid negative effects.

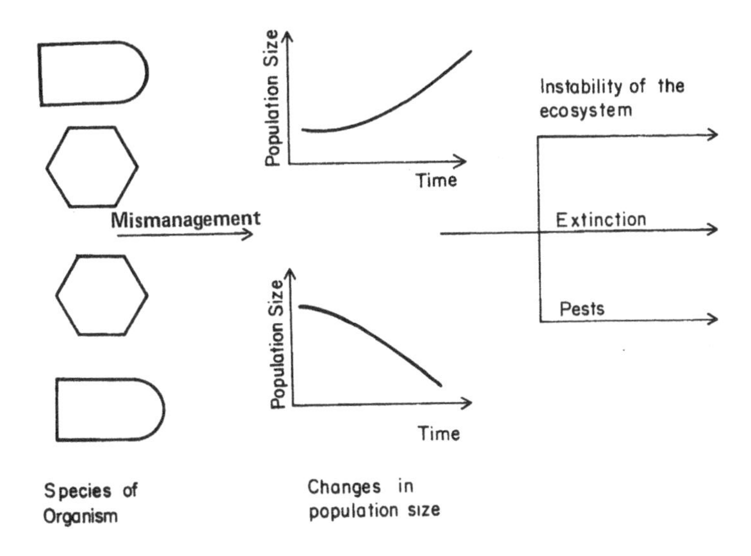

FIGURE 3, The relationship between species diversity and mismanagement.

IV. ECOLOGICAL FLOW CHAINS

To understand the functioning of the desert as a whole is the ultimate goal of the ecologist. In spite of the enormous scientific effort in ecosystem studies, we are not yet able to construct predictive models that describe the behavior of the system. This is because of the diversity of species, life history strategies and the complex relationship between abiotic and biotic elements of the system. However, a possible step in the direction of such a predictive model is supplied by the method of ecological flow chains.

In order to deal with the complexity of the desert ecosystem, one can introduce the concept of ecological flow chains. An ecological flow chain is defined as any flow of matter or energy in the ecosystem in which biotic and abiotic state variables are involved.

In our studies, attention has been focused on energy and soil flow chains. The water flow chain was studied by various authors.[8,9,10] At an experimental watershed covering some 11,325 m^2 with 21 raingauges, it was found that rainfall distribution during a given storm is not uniform because of slope orientation and wind direction. The significance of this finding is that the essential factor that directly controls runoff and erosion processes and indirectly controls - through soil moisture - the spatial distribution of plants and animals, has a non-uniform spatial distribution at any given rainfall. In the same area it was found, that because of systematic spatial variation of the topographic slope properties and surface characteristics, the other two important components of the water flow chain - runoff, and soil moisture - are non-uniformly distributed. Figure 4 illustrates the spatial distribution of the water flow chain of a given storm on a hillside slope.

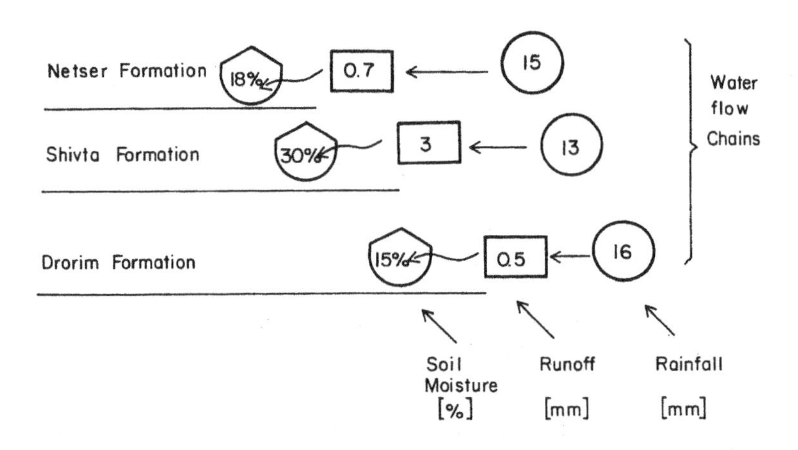

FIGURE 4, Spatial distribution of rainfall, runoff and soil moisture on a hillside slope.

Therefore, for management purposes, we have to look at the water flow chain not just as a simple flow of rainfall, runoff, soil moisture and evaporation, but rather, as a complex network of water flow as shown in Figure 5.

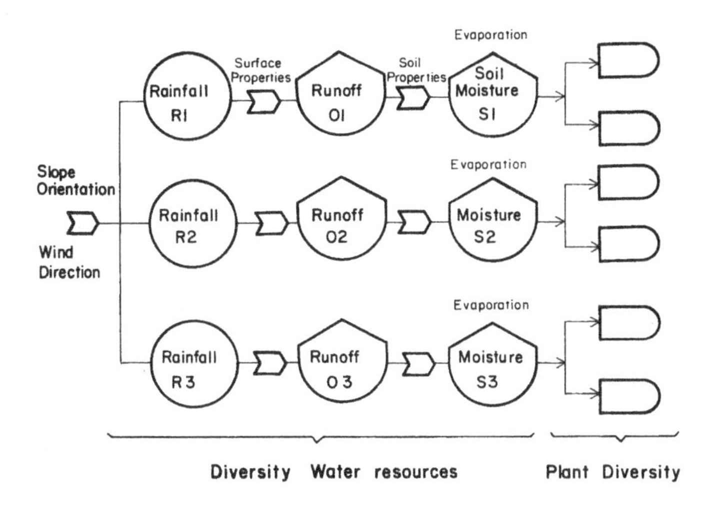

FIGURE 5, The relationship between water flow chains and plant diversity.

Water and solar radiation are the driving forces of the energy flow chain which convert solar energy into chemical energy. The diversity and production of food chains in the desert is highly dependent on the diversity of water resources. As can be seen in Figure 6, when soil moisture is the main source of water, production is much higher than when dew or air humidity provide the water. Thus, it can be concluded that high production in the Negev desert is dependent on soil moisture as the main water resource. Therefore, in the management of the desert, special attention should be given to soil moisture, this being the link between the water and energy flow chains. Soil moisture is obviously related to the presence of soil. Thus, the soil flow chain should be an integral part of the ecological study of the desert. Yair[10] found a strikingly non-uniform contribution of sediment from contiguous plots in a watershed. An analysis of rainfall, runoff and surface properties, such as slope length and gradient, shows that these factors can hardly account for

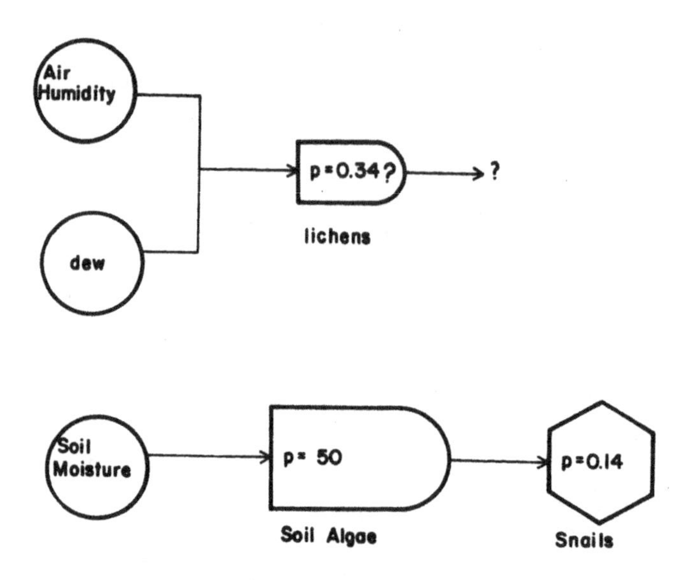

FIGURE 6, The relationship between water resources and production.

spatial non-uniformity in the delivery of sediment. Field observations drew attention to intense digging and burrowing activity by desert animals such as porcupines and isopods. Digging by porcupines, seeking bulbs for their nourishment, breaks up the soil crust which, due to its mechanical properties and biological cover of soil lichens, inhibits soil erosion. Thus, fine soil particles with loose small aggregates are made available for transport by shallow flows. Similarly, burrowing by isopods and their soil-composed faeces which disintegrate easily under the impact of raindrops increase the overall runoff sediment. Data obtained show that the amount of available sediment produced by the animals is of the same order of magnitude as that removed from the experimental site during a single rainy season. This led to the conclusion that soil erosion processes in the desert are part of an ecological flow chain. Figure 7 summarizes the inter-relationship among the water, energy and soil ecological flow chains in erosion processes at the Sede Boqer experimental site.

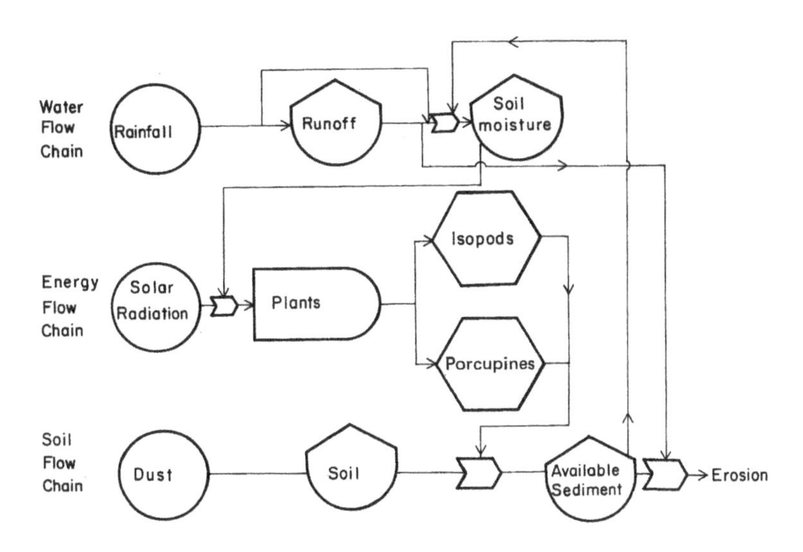

FIGURE 7, Inter-relationship among water, energy and soil ecological flow chains.

This figure illustrates the complex relationship among the three ecological flow chains in which two populations of animals are involved. As was shown before, there are hundreds of species in the desert and a great diversity of landscape and surface properties. It is therefore evident that the desert is a very complex network of ecological flow chains and only scientific investigations by inter-disciplinary teams of meteorologists, geologists, geomorphologists, hydrologists and botanists as well as zoologists may lead to better understanding of the functioning of the desert. With increasing knowledge of the main ecological flow chains in the area, decisions regarding settling the desert can be placed on a much firmer footing.

V. UNSOLVED PROBLEMS

In the foregoing pages we have drawn attention to the enormous richness of species that exist in even small areas of the desert, such as Sede Boqer. We have furthermore used the snail and the isopod to illustrate the great difference that exists in the life history strategies of different species. We have seen moreover, via the use of ecological flow chains, the delicate interrelationships that exist among the various members of the desert ecosystem. Clearly, then, it will not be possible to make quantitative statements about Man's effect on the ecosystem - and *vice versa* - until the life history strategies of all relevant species are known.

In particular, it will be necessary to understand the sensitivity of each species to the presence of Man. One assumes that of all terrestrial ecosystems, the desert ecosystem is one of the most sensitive on account of its unpredictable physical variables (such as rainfall). If this assumption is true, it is of vital importance not to increase this unpredictability by carelessly introducing the additional factor of Man.

We can quantify this need by enumerating six open questions that require the urgent attention of desert ecologists:

1. The relationship between the degree of species diversity and stability in desert ecosystems.
2. Man-made variations in the environment and their relationship to natural species stability.
3. Man-made variations and their relationship to fluctuations in the number of organisms.
4. Adaptation of desert organisms to environmental changes.
5. Changes in the life history strategy of organisms in relation to Man-made systems.
6. The relationship between natural and Man-made ecological flow chains.

If this kind of research is not given urgent attention, there is a danger that the confrontation of Man with the desert ecosystem will have unfortunate consequences for both parties.

VI. REFERENCES

1. A.B. Lovins, Environmental Conservation, 3(1),3 (1976).

2. R. MacArthur, Ecology, 36, 533 (1955).

3. J.T. Giesel, Ann. Rev. Ecol. Syst., 7, 57 (1976).

4. M. Shachak, Y. Orr and Y. Steinberger, Israel J. Malac., 5(1-4), 20 (1975).

5. M. Shachak, E.A. Chapman and Y. Steinberger, Oecologia, (Berl.) 24, 57 (1976).

6. M. Shachak, E.A. Chapman and Y. Orr, Israel J. Med.Sci., 12(8), 887 (1976).

7. M. Shachak, Y. Steinberger and Y. Orr, Oecologia (Berl.) 40, 133 (1979).

8. M. Evenari, L. Shanan and N. Tadmor, The Negev - The Challenge of a Desert (Harvard Univ. Press, Cambridge Mass., 1971) p. 345.

9. D. Sharon, Israel J. Earth Sci., 19, 85 (1970).

10. A. Yair, In: Report of the Commission on Present-Day Geomorphological Processes, ed. H. Poser (Vandenhoeck & Ruprecht, Göttingen, 1974) p. 403.

Part Three

THE DESERT AS HUMAN HABITAT

THE DESERT AS HUMAN HABITAT

The final section of this book addresses itself to man himself and his comfort - both spiritual and physical - in the desert. Broadly speaking, there are three problem areas that must be attacked if serious attempts are to be made to settle the world's deserts.

The first of these problems involves integration of the indigenous (and often primitive) desert population into a twentieth century form of life-style. This problem is particularly acute when the geographical areas concerned are small (as they are for example in Israel). Here policy-makers and planners can make serious mistakes if they do not have a full understanding of the life-styles and values of the people involved. In the opening chapter Emanuel Marx discusses some aspects of the life-style of the Bedouin of Sinai.

The second type of problem to which planners must address themselves pertains to the dynamics of city growth in the isolated desert milieu. In this area much is to be learned from a number of development towns that were hastily built in Israel in the fifties and early sixties in order to house immigrants from various parts of the world. Alex Weingrod addresses some of the social aspects of this phenomenon, and Yehuda Gradus and Eliahu Stern present us with a geographical analysis.

Lastly, we come to the all-important question as to how desert settlements of the future should be built. This question is here dealt with at two levels: the macro and the micro. In the former, Arie Rahamimoff discusses the details of planning whole towns suitable for desert climates. At the micro-level, Baruch Givoni addresses himself to the details of building design. Here the idea is to enable settlers of the future to bend the extreme desert environment so as to enable them to enjoy comfortable climatic conditions for little expenditure of that increasingly scarce commodity, energy.

ECONOMIC CHANGE AMONG PASTORAL NOMADS IN THE MIDDLE EAST

EMANUEL MARX

I. INTRODUCTION

This paper discusses two seemingly contradictory trends among pastoral nomads in the Middle East. First, it deals with the effects of the economic development of the region in recent decades. On the one hand this has created numerous employment opportunities for unskilled and semi-skilled labourers. On the other hand, it has brought about new urban and rural development, which has cut into the critically important summer pastures of the nomads. In consequence, many of the smaller herdsmen are seeking employment in urban centres. Wage labour has become the major course of income for these nomads.

Another result of economic development has been a great increase in the price of meat. In response to the demand, individual entrepreneurs among the nomads have begun to raise animals on a large, almost industrial, scale, and in this manner they can cover their considerable overheads and obtain the necessary summer pastures. The tribesman's economy has thus become diversified and transformed.

Second, the paper deals with the persistence of traditional frameworks and economic pursuits in spite of these vast economic changes. One might assume that in these conditions pastoral nomadism as one knew it was doomed. It is true that some nomads leave their tribes for good, that the herds and fields of most tribesmen are neglected, and that the range and frequency of nomadic movements have become very limited. Nevertheless, pastoral nomadism continues, and the tribal framework is maintained intact though both have changed in the process. One finds everywhere that:

1. Nomads who have taken up employment in towns congregate in tribal residential enclaves and tend to concentrate in a limited number of work places;
2. They maintain strong ties with kinsmen who remain in tribal territory and many even keep their homes and families there;

3. even where few of the nomads still rely on pastoralism for a living, corporate descent groups remain intact and tribal territory is protected;

4. They raise small flocks and till land in the tribal territory, even though they may lose money on these ventures.

I shall argue that these two trends are not as contradictory as they might seem. While wage labour outside the tribal area is the major source of cash income, traditional affiliations and economic pursuits are maintained for reasons of security. Economic and political conditions are insecure, and people are aware that they are likely to lose their urban jobs at any time. The dependence on external sources of employment over which no control can be exerted, entails serious risks. These are perhaps magnified by the nomad's conception of an inconstant ever-changing world. The uncertainties of the desert environment make them aware of the precariousness of life in general. They safeguard the tribe, their territory and the indigenous economy, so that they can resume a pastoral way of life at short notice.

Some nomads achieve a large measure of economic security in town and break away from the tribe: they neglect the kinsmen they left behind, their herds and gardens, and even their tribal friends in town. Sometimes even close kinship relations are allowed to lapse. But others, usually the majority, belong to two worlds. Unable to achieve a permanent foothold in the world of wage labour, they make every effort to insure against an expected loss of employment. They maintain a "basic economy" in their tribal area which can be built up into a subsistence base, and maintain a set of multiplex relationships for mutual assistance in their home grounds.

In order to understand the processes, it is necessary to do away with certain widely accepted notions such as these: Under strong government,nomadic pastoralists settle; sedentary nomads become peasants; under the impact of a modern economy tribal organization breaks down. The literature is replete with statements of this type. Here is one example:[1] "Traditional relationships and groups lose ground and are replaced (usually forcibly) by new superordinated systems, and old personal allegiances of members to the 'small group' are replaced by a new anonymous link of the individual to the big society." Such sweeping statements cannot easily be verified, as long as the conditions in which they apply are not specified. But among the pastoral nomads of the Middle East to-day, they do not seem to apply.

II. INTEGRATION IN THE WIDER ECONOMY

The nomadic pastoralists in the Middle East have always maintained exchange relations with differentiated external economies. They raised animals for sale in rural and urban markets, and in return bought farm produce and manufactured goods. Because there were no limits to the number of animals that the market could absorb, the nomads constantly sought to maximize production and considerable differences in welath ensued. Yet they had to maintain egalitarian corporate groups, both for their personal defence and to maintain control over tribal territory. So some of the wealth had to be redistributed, whenever inequalities became so glaring as to jeopardize the functioning of the corporate groups.

During the present century, and especially since the Second World War, the Middle East has changed rapidly. The economy has grown and become diversified and the population of the region has multiplied. This has had serious repercussions on the nomadic pastoralists. They have been absorbed in the external economy. Many of them work outside the tribal area in many types of work, their indigenous economy also becomes differentiated, they acquire modern means of transportation which allow many of them to establish permanent homes. The main reason for all these changes have been the new economic opportunities: They provided attractive alternatives to pastoralism for many of the smaller herdowners, and they also encouraged the larger herdowners to build up their herds.

The intensity of interaction between any aggregate of pastoral nomads and the external economy can be plotted on a scale. At one end, societies such as some East African cattle breeders, produce for their own subsistence and have no access to markets. Then there are pastoralists who exchange their animals and animal products on the market. They are linked with an external economy, and can still select points of contacts with it.[2] This is the traditional market exchange so often described in the literature[3] as relations between Desert and Sown Land. Then a time arrives when the external economy can no longer be kept at bay. This is the other end of the scale where nomadic societies have become fully integrated in the wider economy. This may assume a variety of forms, yet in the Middle East today these appear to include neither the transformation of pastoralists into peasants, nor the breakdown of corporate organizations and tribal affiliations. Everywhere the pastoralists are reported to be working in towns or industrial plants. Reports also indicate that corporate and tribal allegiances are maintained. While the tribesman earns his living in town, his family often remains in the tribal area and migrates with the herds and may continue to live in tents. Or the family may remain in tents, but give up animal husbandry. It may also move into more permanent dwellings, such as mud huts or stone houses, and yet keep animals. Some examples will be discussed below.

The shift from market exchange to encapsulation in the wider economy is far-reaching in its effects, but they are less destructive of "traditional" organization than one might expect. That is due to the insecurity of much of the work available to the pastoralists in urban conditions. Not that the pastoral economy is all that secure. Droughts, epidemics, floods and raids can destroy a man's wealth almost overnight. But tribal life provides two separate and partly overlapping systems of mutual assurance, agnation and kinship, as well as common access to specified natural resources found in tribal territory. These are valuable assets, which the pastoralist wishes to safeguard even when he earns his living in urban pursuits. Henceforth, the main purpose of tribal life is, for him, to maintain these connections.

This development cannot be explained as a transition from a "primitive" to a "modern" economy, from security due to strict frugality to the insecurity due to unlimited wants.[4] For the pastoral nomads of the Middle East have always been competative producers and also developed some types of conspicuous consumption, such as keeping horses and the exercise of lavish hospitality. In this sense their outlook has always been "modern". Insecurity is simply immanent in most urban pursuits, as will be shown. The shift of the pastoral nomads to urban diversified employment has not received enough attention in anthropological literature. As far as I am aware, there exists only one monograph that examines this recent trend in detail.[5] On the other hand, frequent reference to it is made in other works.[6,7] The limited coverage may be influenced by the continuing concern of the nomads with pastoralism. They often view economic diversification as the end of tribal life, wage labour outside the tribe as a temporary occupation suitable for young men, and pastoralism as the major occupation and the main source of income of the tribe. The field worker too is tempted to concentrate his attention on the well-integrated culture of the pastoralists, the "real" tribesmen. He finds it hard to follow the migrations of numerous men of one tribal unit dispersed in various workplaces in different locations. He is reluctant to view the dilapidated shanty towns as social units that deserve to be studied in their own right. Yet it is to be hoped that once the permanence of the economic change and the significance of wage labour and shanty towns are fully appreciated, more attention will be paid to their study. Only then the realities of life in the tribal area will be seen in a new light.

Typically, the poorer and younger unmarried men seek work outside the tribal area. Even the menial temporary employment they obtain is better than being dependent on their kinsmen at home. When more lucrative work becomes available, those with property and families also go out until sometimes most of the male working population is affected. That is the case in South Sinai today. These nomads possess few skills that townspeople can use, and so inevitably they enter the lowest ranks of unskilled work. they become domestics, farmhands, construction workers or watchmen, often in small establishments, and are paid less than the going rates. In most cases, the workers take care of their own food and lodging. Judging from their own accounts, the nomads may spend several years in the service of one employer, but however long they stay, they are always liable to be dismissed at short notice. They are well aware of their insecure tenure and are always ready to change jobs when required. Yet they also know that they are earning more than they would in the tribal area.

After working some years in town, they acquire a limited range of skills and other experience. They learn how to deal with employers and find alternative jobs. Their income usually rises and some men obtain reasonably secure employment. These men may reach a point where they ask themselves whether they should remain permanently in town and bring their families there, or whether they should terminate their urban employment after a while, and leave their families back in the tribal territory. Whatever the decision, it is usually reversible. And whatever it is, it does not contradict the tendency for the town migrants to maintain links with the tribe. Most of them opt for keeping their families in tribal territory, but even those who move their families to town do not usually relinquish the tribal links. Thus about half the members of the 'Aleqat tribe of South Sinai have settled in Egypt proper with their families, yet ties of kinship are constantly renewed by marriages with tribesmen from Sinai and by mutual visits. Only a few men sever their ties with the tribe, and the best indicator of such intentions is when they cease sending remittances to their families back home. All others try to live in two complementary worlds: they become townsmen whose roots are still in the countryside.

The move to town is facilitated by kinsmen who have preceded the new migrant. They provide hospitality during the first months, and often the kinsmen remain together for years. At the least, they try to obtain lodgings in the same neighbourhood. The old established migrant also finds the newcomer a first place of work. This is likely to be with their own firm or in the same area or type of work. Thus the newcomer remains close to kin and tribesmen who exercise considerable social control over him. Only the dynamics of the urban economy gradually lead to a wider distribution of the tribesmen. Employment vacillates seasonally or cyclically, workplaces expand or close down, and workers may accumulate skills and some capital. And as each man in the course of his duties makes numerous acquaintances in town, he may move from one place of employment to another, and gradually drift away from fellow-tribesmen. Yet the insecurity is so great that most of these mobile people stay in the old residential neighbourhood, and in their turn try to bring kinsmen from back home to their new places of work.

When these townspeople acquire wealth, they invest it in either polygynous marriages, in jewelry for their wives, or in automotive machinery, such as trucks, pick-ups and tractors. All these items appear to have a common denominator: they protect their owners against expected political or economic change. Wives bear children, and the more sons a man has the wider he spreads the risks of future unemployment. Both jewelry and automotive equipment are portable; should conditions in the place become unsuitable, they can be moved, or assist in the move, to a new location. Yet in one respect, the nomads do not behave like other insecure populations: they do not invest in formal education and professional training. While they are aware that most professional qualifications are mobile commodities, they lack the necessary headstart. Their limited educational background - many of them are illiterate - effectively prevents this. Even their children cannot usually make sufficient progress. There are of course exceptions, such as the Bedouin boy who became a university lecturer.[8] It often turns out that highly educated Bedouin are the sons of the wealthier members of the tribe, who themselves were literate.

Kressel[5] examines an urban community of Bedouin from the Negev in the Tel-Aviv metropolitan region. The hamlet of Jawarish was set up by the state in 1952 for a group of Bedouin. Until the early sixties, the population remained stable. A military administration had tried till then to confine the Negev Bedouin in a reservation, where they would pursue the traditional herding and cultivation. Those were days of severe unemployment, and in this manner the Bedouin were kept out of the labour market. By the early sixties the situation had changed. There was full employment and the restrictions on movement were lifted.[9] But when the Bedouin flocked to the centres of employment, only the lowliest manual jobs were left to them. They became seasonal farm hands and worked in other unskilled temporary jobs. These were often obtained without the mediation of the Labour Exchange, and thus illegal. In 1968 work-hungry Arabs from the recently occupied Gaza area and the West Bank of Jordan moved in and undercut the local Bedouin whose income had by then risen considerably. As a result, the employment structure of the people of Jawarish was transformed. A number of them became mediators of work for Arabs from the occupied areas, others bought trucks and tractors and took over the cultivation and harvesting of Jewish farms. Some began to farm on their own account, making use of the cheap labour. Again most of this work was wrought with risks, either because it infringed the prerogatives of the Labour Exchange, or because the land for cultivation could only be obtained by devious means. Insecurity there was, but at the same time capital and skills accumulated, which made it easier for the Bedouin to adapt to further changes. The trend towards economic diversification was unmistakable.

The increased prosperity in the late sixties resulted, among other things, in a sharp rise of polygyny.[5] Kressel argues that this is a peculiarly Arab type of conspicuous consumption. But another interpretation fits his data: namely, that Jawarish men marry a second wife and assume the attendant increased household responsibilities because they wish to raise more progeny. In this way they invest in their future economic security, and not in conspicuous spending.

Cole's short discussion of new economic developments among the Al Murrah Bedouin of Saudi Arabia points in the same direction.[6] "Many tribespeople have settled in shanty town complexes...At least three such shanty towns belong to and are predominantly inhabited by members of the Al Murrah... all of them have settled because of their activities in industrial occupations". The settlers also operate trucks and taxis. Only few of them engage in farming. The oil industry appears to have provided so many opportunities of lucrative employment that the Bedouin are drawn towards its centres. Yet here too, they do not simply disperse in towns, but stay close to their fellow-tribesmen and indicate in other ways that they do not wish to forego the security of the tribal framework.

Economic diversification takes place in tribal territory as well. Herding and farming are no longer the main sources of income of all the tribesmen. In many areas, wage labour becomes the most important occupation, and the number of full-time pastoralists declines. But some people now become pastoralists on an industrial scale. They maintain herds of sheep and goats - camels become less important - amounting to hundreds and even thousands, and employ other tribesmen as herders. Only such a large-scale operation can successfully overcome the obstacles set by the expanding urban and rural settlements. The critical late summer pastures are nearly always located in well-watered, and therefore settled, areas, and pasture rights now are negotiated with the owners, either government or big landowners. But once such arrangements are made, the animals can graze most ot the year in areas where there is plenty of water and pasture, and no longer need to return to the distant desert. Thus many of the large Bedouin flocks of the Negev now stay all year round in the area around Tel Aviv. The Bedouin cannot lease waste land from the government, and overcome this difficulty by joining up with Jewish partners. Sometimes the Jew leases government land, and the Bedouin provides the flock and shepherd, and the two share the profits according to an agreed formula. In other cases, the Bedouin signs an agreement with a Jewish village, which may include additional clauses; for instance, the Bedouin may guard the settlement's orchards.

Once again, Cole's material seems to validate this argument. He argues[6] that fewer people now engage in herding but that the overall number of animals has not declined. Sheep and goats have replaced the camel, because of the insatiable demand for mutton among the urban population. Many herdsmen now spend the summer camped at a permanent well, and migrate during the rest of the year from one pasture to the next. Water is brought to the animals by truck, so that they can range in the desert at some distance from the deep permanent wells. The herds may even stay all year round outside the tribal territory. One may assume that the same trucks which supply water to the flocks also eventaully haul the animals to the urban markets. The organization of herding has thus altered radically.

To sum up the main argument: In recent years, a fast-growing external economy has offered the poorer tribesmen attractive work of various types outside the tribal area. The tribesmen's major source of income now, is wage labour. The labour migrants reduce their investment in pastoralism and cultivation, but as I shall show, do not altogether neglect them, and also maintain their tribal affiliation. Some of the wealthier tribesmen have developed an industrialized type of pastoralism around permanent wells. The income of the tribesmen from all these sources has increased, but they consider them to be less secure than their traditional mode of pastoralism and cultivation. They trust in their tribal affiliations. Their efforts to maintain a secure base in the tribe will be described in the following section.

III. MAINTAINING TRADITIONAL FRAMEWORKS

Many pastoral nomads obtain most of their income by activities in the external economy. But they do not rely on it in the long term. They are keenly aware of the political and economic factors affecting it, and know that upheavals of any kind adversely affect people on the bottom rung, like themselves. A Bedouin construction worker in Sharm el-Sheikh, South Sinai, spoke quite explicity about two kinds of insecurity to which every Bedouin was exposed: "Any Jewish workman can throw you out of work, even if you have been three years on the job and he arrived only just now. We are not safe here; what will happen if there should be another war? When the October (1973) War broke out, we were left stranded here and walked three days in the mountains without food or water before reaching home. In the mountains we are safe, so we prefer our families to remain there".[10]

Similarly argued a Bedouin of the Saudi Arabian Al-Murrah, many of whom work in the oil industry: "The oil wells can be blown up in thirty minutes and, with no money, all those people in Dhahran and Riyadh would die from lack of food. Why, they would not even have enough gasoline to leave and go back to their homelands".[6]

Both speakers implied that back home they had a secure economic base that would at least feed them and their families. I claim that much of the nomad's tribal economy and social organization, in the tribal area and in town, can only be understood in these terms. Tribesmen stick to their traditional economic pursuits even when they earn good wages outside the tribe, and even when their flocks and gardens do not yield profits. They do so in the knowledge that in times of need they can fall back on their traditional economy, and by devoting all their efforts to it could make it work again. Their subsistence would thus be assured. It is for related reasons that they foster their kin and agnatic relationships and maintain the tribe. Even the tribal neighbourhoods in towns and the tribal shanty towns serve this purpose. Taken together, these economic and organizational patterns indicate how deeply the nomad is concerned about providing a secure livelihood for his family.

I shall now examine the various elements of this "secure economic base" one by one. First, I discuss the tribal enclaves set up in towns. In the Negev, Bedouin settled in Jawarish again provide the best illustration. They preserve their tribal identity; they reside in residential clusters, co-operate as groups in local politics, and maintain visiting and marital links with their groups of origin. In a violent local dispute they call on their agnates from the Negev for help, and in the end the quarrel is patched up by their respective tribal chiefs from the Negev.[5] The local associations of migrants, here as elsewhere, serve numerous ends; they help the new arrival to find his bearings, and provide mutual assistance. It is significant, however, that among these former nomads the associations take the form of offshoots of agnatic descent groups in the area of origin. The links with the groups in the Negev can be useful in two ways; first, members of the group can be enlisted for help in disputes (that is of course a two-way traffice); and second, the line of retreat to the Negev is always kept open. Thus, whenever there is a threat of war between Israel and her Arab neighbours, residents of Jawarish return to the Negev. The view that the Negev is the warm womb to which one can return is held by these Bedouin in spite of the fact that they do not possess a territory. Their tribal area there is largely made up of small parcels of land leased by individuals from government, with annually renewable contracts.

The link between the local groups of migrant labourers and their tribe is even more pronounced among nomads who actually possess a tribal territory. Here there is constant movement between the town and the tribal base, and the stay in town is viewed by many of the migrants as a temporary episode. Among the Bedouin of South Sinai this is expressed mainly in two ways. First, many migrants do not set up house in town; they lodge either with other tribesmen, or a number of them jointly rent cheap lodgings, or they stay overnight in their place of work. Often they do not possess so much as a corner where they can leave their few belongings. Second, they do not usually bring their families to town. Their homes are then clearly in the tribal area. This pattern of behaviour was observed in such urban centres as Eilat and Sharm el-Sheikh.

The Al-Murrah seem to stand halfway between Jawarish and South Sinai Bedouin. The wage labourers of this tribe reside in shanty towns located on the fringe of cities, among members of their tribe. "There is no village or town organization and a man looks to his nomadic tribal leader and would plan to intermarry with his nomadic kinspeople rather than with other families in the settlement".[6] Yet the migrants do build houses. At first they are constructed of scrap materials, but each year some improvements, "such as concrete floors and plywood walls inside the rooms"[6] are introduced. Then gardens are planted, and small flocks raised, and the urban settlement seems to consolidate. In spite of all this, the ties with the tribal territory are faithfully maintained. "About half the households keep tents, and their families, especially the women and younger sons, move out to graze their sheep and goats in the desert during part of the winter and spring. During summer vacation, most of the young sons and many of the women enjoy a visit with their camel-owning relatives at their wells deep in the desert".[6]

Up to now the discussion has centered on the settlements of wage labourers. Therefore the illusion that the tribal economy and organization was sound and healthy could be maintained. How else could one explain the labour migrants' view that their home and political centre, as well as their means of economic survival, were located out there in tribal territory? But when the focus of attention shifts to the tribe another picture emerges. It turns out that the traditional economic pursuits of the tribesmen are neglected; one is told that the land is deteriorating and can no longer sustain the flocks, that the palm trees are not profitable and that the people depend on the income from wage labour. Not only the tribal economy seems to be on its last legs, but the tribe as well. Most of the men are away at work and return for short visits. Only the old men, women and children stay at home. All this is, of course, true, but must be seen in a different perspective. For what one sees is not a slow deterioration of the tribe, but rather a large-scale maintenance operation. This involves three main interrelated aspects: The maintaining of kinship ties, of agnatic descent groups and tribal affiliation, and of a "traditional" economy. For most of the year these activities are carried out by the people back home, but their successful accomplishment requires the periodic co-operation of the wage labourers. They must return for defined periods and join in the activities. Thus the secure home base is kept ready for reactivation in time of need. As in so much of human endeavour, when times are good, the maintenance operations are carried out half-heartedly. Only when bad times are envisaged, they are carried out in earnest and their full significance emerges.

This argument can be illustrated by observations made among Jebaliyah Bedouin of South Sinai. They live in the high mountains in the centre of the peninsula, a harsh desert that permits horticulture only in a few oases and in the mountain valleys. Grains and most of the other basic foods are imported and not even dates are produced in sufficient quantity to satisfy local demand. Bedouin have acquired a few specialized skills; there are some builders, well-diggers, truck-drivers, mechanics and some other artisans and a handful of shopkeepers. But wherever one moves in the area, one encounters women and girls herding flocks of black goats, with here and there some sheep interspersed. These flocks vary in size from a few head to fifty and more. They are so ubiquitous, that one may easily gain the impression that pastoralism is the chief source of income for these people, and also that here for once women are the main providers. In fact, herding is practised on a limited scale, mainly because Bedouin consider that they either make no profit or actually sustain losses from it. The average Bedouin owns an estimated 5-6 goats and one camel. Yet it is an indisputable fact that many Bedouin households raise animals.

Horticulture is even more important for the Jebaliyah, as it can more easily be expanded, and because it does, even in these days of wage labour, provide some of the Bedouin's subsistence. I shall discuss it below in somewhat greater detail. Practically all the adult males of the tribe are engaged in wage labour. They tend to stay at work for two to three weeks at a time, without rest-days, and then to return home for a short, often unspecified, period of rest, which often becomes longer than initially planned. During their home leave, they visit relatives and friends, who in their absence have taken care of their families and economic affairs. The Bedouin generally maintain joint households, into which go the labour migrant's wages. Joint households are often maintained by the families of brothers, but households of brothers-in-law or cousins are also frequently found. The extent of co-operation varies but some of the joint households arrange and finance the marriages of members, nurse members through long illness, and provide for widows and orphans. These groups may co-operate for long periods, and where that is the case they provide members with basic security in case of death, illness and poverty. Joint households are, of course, found outside the desert as well, but they can thrive only where economic survival is assured. As long as Bedouin labour migrants must fear wholesale dismissals from work, perhaps in the wake of a political upheaval, the joint households must remain in the mountains. There gardens and flocks guarantee at least some income, and the members can support each other for an extended period. Here physical proximity is of the essence. The children of migrants can be properly cared for where they live next door. Their flock will be taken to pasture in turns by the girls of the other members of the household, only if it is kept with the other animals. Provisions for the migrant's family can be bought and stored when they live nearby. There is thus a tendency for members of joint households, and of other kin, to settle in a cluster, usually near a place where food and services are available. This arrangement reduces the mobility of the individual households. It also affects the small flock, which no longer determines the migrations of the households. It is taken to pasture in the vicinity of the hamlet, where the food is soon exhausted. The animal's diet, has to be supplemented for about six months every year by feeding them expensive imported corn. No wonder the Bedouin claim that they lose money on their flocks; they therefore try to keep them as small as they can, but they do not abandon them.

Tribal membership is still important for the pastoral nomad, because it gives him the right to exploit certain resources within tribal territory, and others outside it. The individual acquires membership of the tribe indirectly, by belonging to a patronymic group or a patrilineal descent group which is recognised as part of the tribe. The outward sign of the group's belonging is usually a place in the tribal genealogy. The small corporate groups of agnates which are characteristic of the pastoral nomads in the Middle East have little work to do in today's conditions. Cultivable land is generally owned by individuals, and pasture by the tribe; blood disputes are not often prosecuted by these groups; self-help and vengeance are frowned upon by police forces. So the remaining purpose of the agnatic descent groups is to mediate the individual's membership of the tribe.

The tribe reserves for its members certain rights over the strategic resources found in its territory, such as water for the irrigation of gardens, housebuilding sites, pasture and employment opportunities, or any other combination of rights. Due to the extensive concern with wage labour, some of these rights may not be fully exploited by tribesmen. There may be a justified fear that these rights could lapse. Therefore the tribesmen devote considerable efforts to the presevation of the tribe. In the absence of joint activities, they can do so chiefly by organizing gatherings, at which tribal solidarity is reaffirmed. This annual meet is set to coincide with the date harvest or the ripening of other produce, so as to attract as many people as possible. Many of the labour migrants leave their employment at such times and return to the tribal area. In South Sinai, hundreds of tribesmen congregate in the major oases of Dahab, Nuweba and Wadi Firan, and in some smaller ones, and at the end of the date harvest the gatherings culminate in tribal pilgrimages, each tribe gathering at the tomb of a saint.[10] Among the Al-Murrah there are the beginnings of such tribal gatherings. Cole states that the four oases in the tribe's territory now "play an important role in the maintenance of tribal solidarity... Since the advent of wage labour... many (tribesmen) have begun to invest some of their income in the construction of summer homes in the oases... Those who are engaged in full-time wage labour return during summer vacations to the oases where they have a chance to renew their relationships with the rest of their lineage and clan".[6]

I have made some reference to the nomads maintenance of traditional economic pursuits, again in order to have something to fall back on. An example that has been studied in some detail is the horticulture of the Jebaliya of Mount Sinai. The Jebaliyah are the second largest tribe in South Sinai, with about 1200 members. Practically every household owns one or more orchards in the mountain valleys, where they grow fruit trees, like apricots, apples, pears, almonds quinces and pomegranates and some vegetables. Each orchard has its own well and is surrounded by a dry stone wall. These orchards could supply the Jebaliyah's basic food requirements. Up to the 1950's many households lived on the produce of their orchards. Part they consumed, and part, especially a robust variety of pear, they transported on camel-back to urban centres and with the proceeds were able to buy enough grain to last them through the year.

During periods of abundant wage labour the Jebaliyah view their gardens as pleasant summer retreats. They spend part of the hot summer months in the shade of the trees. The gardens are watered fairly regularly, and from time to time men carry out some maintenance. But they consider neither the fruit nor the vegetables as valuable economic resources. They are viewed as delicious additions to the diet, but not as subsistence. For wage labour has brought them higher incomes, and reduced the relative value of the garden produce. Yet they do tend the orchards, so that they bear fruit. If necessary, they would again rely on the proceeds of the orchards, just as they did in the old days.

Such a contingency actually arose after the October 1973 War between Egypt and Israel, when most Israeli economic activities in South Sinai were interrupted for about five months. During that period the Bedouin used up their money and food. Most families, it turned out, had stored basic foods, such as wheat, sugar and oil for such an eventuality. People who had large stocks of food shared them with kinsmen. As everyone was back home, social relationships were intensified. Soon the interest in orchards also increased. Some men planned to acquire orchards or to plant new ones, and to improve existing ones. Tribal pilgrimages in 1974 were very well attended. This flurry of activity was part of the Bedouin's attempt to reactivate their secure economic base. Not everyone was fully prepared for the sudden change, but all the Bedouin shared in the mutual assurance provided by a social and economic organization designed to serve this purpose.

IV. CONCLUSIONS

The behaviour of the pastoral nomads in wage labour and in the tribe is then all of one piece. If for purposes of exposition I talked about two types of economy, wage labour and the traditional economic base, this distinction can now be dropped. What is clear, however, is that the nomads seek to strike a balance between their wish to maximize money income and their wish to provide full social security for their families.

V. ACKNOWLEDGEMENTS

The paper is partly based on fieldwork in South Sinai, carried out from 1971 onwards. This work was generously supported by the Ford Foundation, through the Israel Foundation Trustees. An earlier version was presented at a Conference of the Commission on Nomadic Peoples of the International Union of Anthropological and Ethnological Sciences, in London, in 1978. I gratefully acknowledge comments received from S. Hartman, G. Kressel and S. Weir.

VI. REFERENCES

1. Ferdinand Scholz. Belutschistan (Goltze, Gottingen 1974).

2. Paul T.W. Baxter in Pastoralism in Tropical Africa ed. T. Monod, (International African Institute, London, 1975).

3. Raphael Patai, Golden River to Golden Road: Society, Culture and Change in the Middle East, 3rd ed. (Univ. of Pennsylvania Press, Philadelphia 1969).

   Fredrick Barth in The Desert and the Sown ed. C. Nelson, (Univ. of California, Inst. of International Studies, Berkeley 1973).

   Alois Musil, The Manners and Customs of the Rwala Bedouins (American Geographical Soc. New York, 1928).

   Warren W. Swidler in The Desert and the Sown ed. C. Nelson (*op. cit)*

4. Marshall D. Sahlins, Stone Age Economics, (Tavistock, London, 1974).

5. Gideon M. Kressel, Individuality Against Tribality: The Dynamics of a Bedouin Community in a Process of Urbanization, Hebrew, (Hakibbutz Hameuchad, Tel Aviv 1976).

6. Donald P. Cole. Nomads of the Nomads: The Al Murrah Bedouin of the Empty Quarter (Aldine, Chicago 1975).

7. Ahmed M. Abou-Zeid in Mediterranean Countrymen ed. J. Pitt-Rivers (Mouton, Paris 1963).

   Abdulla S. Bujra in The Desert and the Sown ed. C. Nelson, (*op. cit)*.

   Fidelity Lancaster, New Society 591, 245 (1974)

   Gillian Lewando-Hundt, Women's Power and Settlement, thesis (Univ. of Edinburgh, 1978)

8. Isaak Diqs, A Bedouin Boyhood, (Allen and Unwin, London 1967).

9. Emanuel Marx, Bedouin of the Negev (Manchester Univ. Press, Manchester, 1967).

10. Emanuel Marx in Regional Cults ed. R.P. Werbner, (Academic Press, London, 1977)

DESERT TOWNS AS A SOCIAL TYPE

ALEX WEINGROD

I. IDENTIFICATION OF THE PROBLEM

Reading the literature on the social features of desert towns often seems as barren and unpromising as the deserts themselves. The descriptive literature is meagre - a short note on an oasis town here, the depiction of rapid growth in a desert town there - and the few attempts at generalization tend to be cryptic and partial.[1] There are no full scale social scientific or historical studies of the formation, growth and development of towns in the desert, and consequently it is hardly surprising that a convincing model of desert town formation has yet to be presented. Moreover, there are at least two basic problems involved in the construction of such a model. First, the range of cities and towns to be included under the heading of "desert" can be bewilderingly large; for example, in his analysis of urban growth and manufacturing in what are called "arid lands" Andrew Wilson[2] includes such diverse cities as Cairo, Fresno, Los Angeles, Riyadh and Cuzco. The obvious question to ask is how one can draw meaningful conclusions from a listing of cities whose histories, civilizations and populations are so thoroughly different. Second, attempts to consider the "social features" or "social issues" typical of desert settlements often end up discussing processes and issues that are characteristic of all cities. That such general processes as rapid population growth or marked social stratification are characteristic of some desert towns is hardly surprising; change and heterogeneity are, after all, among the defining features of urban life. The question (and it is by no means an easy question) is whether there may be additional features that pertain uniquely to towns in the desert.

These obstacles may appear as grim as the desert itself - and yet, once again like the desert, one has the feeling that there is "something promising out there." The problem is theoretically attractive; can one formulate a meaningful set of statements regarding desert towns as a structural type? Moreover, this is an eminently practical issue for economic planners and those who frame government social and economic policies. Both desertification and reclaiming desert regions are matters of world wide importance, and it is therefore vital to formulate development strategies based upon informed empirical studies. It is, hence, to this difficult yet potentially rewarding topic that I turn first.

II. A TENTATIVE MODEL

David Amiran's essay entitled "Problems and Implications in the Development of the Arid Lands"[3] can serve as a useful starting point. Amiran's interest in this essay is not focused exclusively upon desert cities, yet his comments provide a beginning framework for the analysis of urban development in desert regions.

"First and foremost" he writes[3] "arid lands are poor in water.... As a result of these conditions of the natural environment, arid lands generally are sparsely populated... The population fabric of arid lands is generally discontinuous and brings about relatively strong concentrations of population at a few centres or in a few areas. The result is that both at the traditional and at the modern technological level, urban settlements in arid areas contain a large proportion of the total population... It appears that the limited range of economic development choices available... makes arid zone towns single purpose towns from the functional point-of-view... A subsidiary factor is the element of size. Not only is there an upper limit of town size beyond which special services and organizational measures have to be applied to make the large modern metropolitan area liveable, there obviously is, as well, a lower limit below which settlements established as towns cannot continue to function as such."

This lengthy citation[3] contains the key elements in our model. Indeed, following Amiran it is possible to identify five features that may be said to characterize desert cities and urban development in desert zones. First, city formation and subsequent urban growth is dependent upon providing adequate, continuous supplies of water; second, insofar as desert environments lack abundant natural resources, desert towns tend to be small and limited in population; third, from the economic "functional" point-of-view these towns often tend to concentrate upon a single industrial or other endeavour; fourth, as a consequence of limited resources and small populations, desert towns face special problems of providing adequate modern social services; and fifth, desert towns are typified by the absence of a rural hinterland, and this also limits their potential for socio-economic growth and diversification.

The literature that describes desert towns (places such as an oasis in the Sahara, an Australian mining town, or Phoenix, Arizona several decades ago) tend to support these generalizations.[4] To be sure, some of them (such as the problems of small size or concentration upon a single industry) are not specific to desert conditions and typify cities and city life anywhere. Nevertheless, taken together these five features provide a model in terms of which desert towns can be meaningfully analyzed. At the same time, however, the features require amplification, and in some instances, substantial qualification.

It is clear that scarcity of water and the absence of a multiple resource base are crucial features of our model. Water supplies adequate for human and industrial needs must be provided in dependable quantities and at a price that is economically feasible; thus, to cite one example, the recent population boom in the American South West is based upon transporting large quantities of water over long distances.[5] As regards resources, towns in the desert normally were formed or founded as mining centres, garrison towns, small trade and caravan towns, or in some instances, religious or cult centres. This was as true in antiquity as it is today; Nabatean settlements in the Negev desert were sited there since they dominated ancient trade routes, just as former oasis or new towns in the Sahara or Namib desert are founded upon nearby oil or mineral deposits.[6] To cite still another example, the early development of places such as Phoenix, Arizona, or some of the urban centres in Southern California, was closely linked with large scale investments in nearby military air and naval bases.

The resource base in these desert towns is not broad or powerful enough to support large populations. Mining in the Namib, Australia or the Sahara, as well as garrison and caravan centres wherever they are located, do not require large numbers of persons, and hence these towns tend to have relatively small populations with not much of a base for sustained growth. Moreover, it is this concentration upon a single resource that lends the well-known "boom and bust" cycle to desert towns. Towns based upon mineral extraction often have short-lived periods of prosperity and boom, followed by much longer periods of bust and, at times, almost total collapse. Similarly, garrison centres depend for their continuity upon national strategies and decisions, and as these change, the centres may also recede into the "ghost town" category.

However, as this brief discussion suggests, it may be useful to distinguish between two kinds of desert town: first, those based upon nearby economic resources, and second, those others whose establishment and growth is based upon external national military or other interests. Towns in the second category are clearly not dependent upon the presence of mineral or other resources - whether they grow or decline in population is a function of political decisions taken at the national level. This is a critical point, since it means that under certain circumstances the economic base or a part of it may be externally provided by means of various forms of subsidization (military payements, welfare payments, tax rebates and the like). Desert towns may, in other words, be primarily artificial creations, formed, supported and maintained by national level interests.

This leads to a second point regarding resources and urban growth. The assumption which underlies the analysis thus far is that deserts are such grim, forbidding places that they become settled and develop _only_ if there is some especially attractive resource to exploit. Why should one choose to live in a hot, brown desert environment unless there are clear economic gains to be made? The new and older industrial and mineral-based towns in Australian and North African deserts appear to fit this logic. On the other hand, it does not explain the recent explosive growth of cities in the American Southwest. Places like Phoenix and Tucson have lately become "magnet cities", attracting both new populations as well as new industries on a very substantial scale. Under certain conditions, in other words, desert towns become urban centres equal to those in other ecological zones.

What are these conditions? Why do desert environments suddenly become attractive?

These are difficult questions, much too complex to be adequately considered in this brief article. A number of factors may, however, be suggested. First, the sheer growth of population in "more favoured" zones may at some point make the desert attractive; in the United States example, when such huge population aggregates as the New York or Los Angeles metropolitan regions appear to be bursting with people, the underpopulated desert appears as a zone of new opportunity. The sheer availability of open spaces for industry and housing is another feature. More important still is a deeper shift in perspective: this recent attractiveness of desert cities is part of the new culture that accompanies the transition to post-industrial civilization. Given the new climate-control technology of post-industrial life (home and automobile air conditioning on a vast scale) the desert heat becomes less of an obstacle. Similarly, with the advent of personal wealth that signals post-industrial times, the desert locale can become especially attractive to certain large populations - older persons in search of warm climates for retirement, or those who suffer from diseases that the desert sun and air may heal. Moreover, the desert's appeal may also attract younger persons in search of clean new environments - today's "ecologists" - or those who are simply fascinated by the desert's beauty.

Once these attractive processes have begun the desert towns also spawn other forms of economic activity; for example, new industries choose to move to these cities, and commercial activities also grow apace. Desert cities within the orbit of post-industrial civilization may therefore expand their resource base and rapidly grow in previously undreamt of ways.

Let us next turn attention to a different feature of desert towns. The general model proposed earlier stresses the fact that "settlements in arid areas contain large proportions of the total population"[3] For example, the smallish towns in the desert zones of Sudan or Australia include a large fraction of the entire region's population. However, these beginning statements need greater amplification: what is of interest are the patterns of relations between the urban majority and the non-urban minority.

Desert towns are not truely lacking in a social hinterland. Historically, the desert served as a kind of "refuge region" for those small marginal populations that learned to inhabit these barren areas. The growth of desert cities is closely linked with the fate and future of these populations: indeed, there is an interaction between urban centres and desert inhabitants which is systematic and significant for both. It is for this reason that these non-city populations need to be carefully considered in any analysis of desert town development.

Studies of towns in desert regions often make reference to the native non-urban population; these are such fabled groups as the Bushmen in the Kalahari, Aborigine in Australia, Bedouin in Israel's Negev or in Saudi Arabia, Tuareg in the Sahara, Indians in the American Southwest.[7] Many came to the desert as a hideaway - a place so grim and forbidding that one could retreat to it without fear of conquest and exploitation - while others fled from the law and "civilization." In the desert they successfully adapted to an environment of comparative scarcity, and their cultural inventory and orientation was deeply influenced by desert conditions. Typically small sized and scattered, they nonetheless are a vital part of the desert.

The impact of industrialism and city development upon these populations appears almost inevitably to be massive in scale and ultimately destructive. Drawn into industrial society at its lower rungs - as occasional unskilled workers in a variety of dull tasks - these refuge people must undergo yet another deep crisis of adaptation. Despois analysis of the impact of industrialism on the native population of Saharan oasis towns makes this point vividly:[6]

> "Most of the camps are temporary; most of the immigrants were unable to work as expected and were sent back; others could not withstand the continuous effort required and abandoned their new work. These unemployed withdrew to their own oasis but did not wish to take up their former occupations... The youth and the active males abandon agriculture, and the underground wealth of their countries helps but a few families, 15,000 at the most, of about 500,000 who populate the Sahara from the Atlantic Coast to Egypt"

Conclusions such as these are often repeated in other studies of marginal populations; industrial civilization tends to shatter family and other groups and to set off processes of change that often result in apathy, personal destructiveness and crime.[8] Although these issues are most serious for the once marginal peoples, they also raise difficult problems for the urban population as well. Problems of poverty, illness, substandard housing and crime threaten the entire urban fabric, and they are therefore issues of general concern. Desert towns are rarely formed in a social vacuum, and their development has important consequences for both the native and the new population.

III. SOME TESTS OF THE MODEL

The formation of desert towns in Israel provides an opportunity to test the usefulness of the general model. To be sure, Israeli desert towns are comparatively new and small in size; in this regard they do not encompass all the dimensions suggested by the model. However, insofar as their social and economic features have been documented, they offer an interesting case-in-point.

The idea of Jewish pioneer communities, or of settlers "reclaiming the land", reaches back in time a century or more.[9] Pioneering in the form of new agricultural communities was primarily associated with clearing swampy lands in the north or settling various small zones along the Mediterranean littoral. Pioneering in the desert has a much more recent history. The first agricultural settlements in the Negev were established as late as 1944, and the first towns were created roughly ten years later. More specifically, beginning in the early 1950's government planners began to design and construct a string of new towns in the Negev. Beersheva, the ancient Biblical oasis and later Ottoman garrison town, was expanded to become the "Capital of the Negev", and the outlying new towns of Dimona, Yerucham, Mitzpeh Ramon and Arad were literally created in the desert itself.

From the government planning perspective, these new towns were meant to serve three main purposes. First, since sizeable mineral deposits had been located in the region, the towns were positioned close by in order to provide a convenient residence for workers and others. For example, Arad was sited near to mineral deposits located on the Rotem Plain, and workers employed at the Dead Sea works lived mainly in Dimona and Beersheva. Hence the region's natural resources provided a beginning base for industrial development, and the towns grew around these economic cores. Second, the formation and expansion of the desert towns was part of the national programme of dispersing the Jewish population away from the coastal centres and into the peripheries. This was thought to be strategically vital since without settling and thereby "claiming" the desert zones external international pressures might seek to remove them from Israel's political control. The desert town policy had, in other words, a firm foundation in national political interests and plans. Third, the formation of desert towns was also meant to serve military interests. The towns were intended to protect border areas, and some also became military or garrison centres. In brief, in the Negev case, the formation of new towns was linked to national economic and political interests, with the latter often predominating.

Although there are significant variations among them, the overall trend has been for these towns to grow in population. Beersheva in particular has grown explosively from a population of 16,000 in 1954 to nearly 100,000 in 1975. Sited on the desert's edge, Beersheva's present day absolute size has made it a major Israeli population centre; indeed, during the past quarter century, it has grown to become the practically autonomous hub for regional-level economic and administrative activities (banking and commercial offices, government agencies and merchandising). The smaller desert towns of Dimona and Arad have also shown steady population growth; Dimona grew from roughly 5,000 in 1961 to 27,400 in 1975, while Arad, which numbered 1,300 persons in 1965 had grown to 8,800 in 1975. The absolute size of these two towns also appears to provide a solid urban base. On the other hand, the growth in population of both Yerucham and Mitzpe Ramon has been much slower. The former town grew from 1,600 in 1961 to 6,400 in 1975, while the latter climbed from several hundreds in 1961 to 1,750 in 1975. In general, Beersheva and Dimona have become fairly substantial urban centres, while the other three towns are considerably smaller, and one, Mitzpe Ramon, is small indeed.

Turning back to our general model, special attention was given to the economic resource base of desert towns, and more particularly to the distinctions between towns based upon local resources and those sustained by national allocations. Taken as a group, the Negev towns represent a mixture of these two factors, although certain towns have a greater dependence upon outside national support. The Dead Sea works and phosphate deposits, as well as the secondary industries that process the chemicals, provide a substantial natural resource base. Although they have presented a fairly wide source of employment, these capital intensive industries are not economically powerful enough to underpin a city of 100,000 persons plus four other towns with an additional 45,000 or so residents. Government development policies have therefore, aimed at attracting other industries to the region. The forms of "attraction" are many, but they principally involve various forms of tax benefits and other financial inducements. In this fashion, two textile factories were established in Dimona, a modern bottling plant was built in Yerucham, and numerous smaller plants and workshops are spread throughout the area.

In addition, government investments in military installations and advanced research has also been a major economic factor. Indeed, the expectation is that the planned redeployment of army garrisons from the Sinai to the Negev region will add significantly to the towns' population and economic base.

The range of urban services provided in the desert towns is, in general, rather impressive. Each town oversees its local educational system, and while the excellence of the schools is varied in some instances high quality educational systems have been established. Similarly, the formation of a new university and medical centre in Beersheva has greatly expanded the range of modern urban services. To be sure, particularly in the smaller towns, the level of educational and health services leaves much to be desired, and yet this desert region compares fairly well with the older coastal zone.

In the previous discussion, emphasis was given to the linkage between desert towns and their surrounding society, and more specifically to the transformations that may come about in post-industrial times. Have these changes taken place in the Negev? To what extent have the desert towns become attractive, magnet centres?

There may be pockets of "post-industrialism" in Israeli society, but there is not yet the wealth nor the all-encompassing technology for these pockets to expand throughout the entire society. For example, climate control has not spread beyond a few public places and well appointed homes. In this regard, the desert environment has not been altered. On the other hand, however, there are signs that the region is becoming more attractive. In the past, nearly all of the new towns were formed by immigrants who were transported directly to the towns, allocated jobs and housing by government agencies, and encouraged and expected to make a new life in the desert. Many of these immigrants subsequently left for the central regions, and others were sent to take their place.[10] More recently however, the population has stabilized or grown slowly in part as the result of internal migration. Some families have been attracted by employment possibilities or comparatively inexpensive housing, while others have come for health reasons or in search of a better urban environment. The Negev is no longer so forbidding to many Israelis, and although this migration has been small, it may signal the larger waves of the future.

Finally, let me turn briefly to the impact that urban development has had upon the Negev Bedouin population. (The reader is referred to the chapter by Emanuel Marx for fuller information on this subject). From the residential point of view, few Bedouin have thus far settled in Beersheva or the smaller desert towns; most have settled in permanent small enclaves or in the special towns planned for them. However, many Bedouin males have found full or part-time employment in the cities, and nearly all of them make regular use of regional urban services such as hospitals or courts of law. Many Bedouin are, in other words, thoroughly familiar with the urban centres constructed near to them. What is of particular interest is the fact that they appear to be using these resources without at the same time suffering grievous social dislocations or cultural crises. The Bedouin do not appear to be clamped down on the lowest rungs of the society, nor are they under severe pressure to assimilate into Israeli culture. Instead, they have thus far been able to sustain a separate Bedouin social system that makes use of urban social services without at the same time becoming "urban". There is no "Bedouin problem" in the desert towns, and although they lack modern services, the Bedouin settlements appear to retain a certain tough resilience and vitality. Their situation should certainly not be romanticized - there are numerous instances of poverty, and internal social tensions are often explosive - and yet in their adaptation to the new urban environment, the Bedouin seem to have successfully exploited some new opportunities.

IV. FACTORS REQUIRING FURTHER RESEARCH

This essay is a starting point, a preliminary mapping of issues and test of the utility of a desert town model. It may be helpful to suggest, in conclusion, some of the directions that future research in desert towns might take.

First, we clearly need more detailed, historically-oriented studies that document the development of modern desert communities. Work in developing a model cannot proceed much farther without richly detailed studies of this kind. Second, attention can profitably be given to issues of stratification and internal differentiation; that is, if they tend to form and develop around a single economic core, or have the continuing dominant influence of nationally-directed resources, then these communities may evolve distinct social structures. For example, patterns of stratification may differ between communities of this kind and those that have a wider resource base. Moreover, the population selection that draws different types of people - different in regard to age, sex, occupation and other features - will also have an effect upon urban development. Finally, the interactions between desert towns and the indigenous desert people needs to be more carefully explored. For example, what are the cultural, economic or other features that explain the Bedouin adaptation in contrast to that of other desert people? These are, among others, some of the issues and questions that await future study.

V. ACKNOWLEDGEMENTS

The author wishes to express his thanks to David Amiran and Amnon Shinar for helpful comments on this paper.

The population data in section III come from the Israel Central Bureau of Statistics.

VI. REFERENCES

1. See for example the papers by:
W.A. Andrews and Marion Clawson in C.Hodge and C.N. Hodges, Urbanization in the Arid Lands (ICASALS Pubn, Lubbock, 1974)
and those by: L. Gordon and A.V. Kneese in G. Golany, Urban Planning for Arid Zones (Wiley, New York, 1978).

2. A. Wilson in D. Amiran and A. Wilson, Coastal Deserts: Their Natural and Human Environments (Univ. of Arizona Press, Tucson, 1973).

3. D. Amiran in D. Amiran and A. Wilson (*op. cit*)

4. See for example the paper by G. Hale in Hodge and Hodges (*op. cit*) and that of J. Gentilli in Amiran and Wilson (*op.cit*).

5. W. Weismantel in Hodge and Hodges (*op. cit*).

6. J. Despois in Amiran and Wilson (*op. cit*) and R.F. Logan in W. McGinnies and B. Goldman, Arid Lands in Perspective (Univ. of Arizona Press, Tucson, 1969).

7. See for example: R. Lee and I. De Vore, Kalahari Hunters and Gatherers (Harvard Univ.Press, Cambridge, 1976); W. L. Warner, A Black Byzantium (Harpers, New York, 1958); E. Marx, Bedouin of the Negev (Manchester Univ. Press, Manchester, 1967), C. Kluckhohn, The Navaho (Harvard Univ. Press, Cambridge, 1948).

8. E. Spicer, Cycles of Conquest (Univ. of Chicago Press, 1962).

9. D. Weintraub, M. Lissak, Y. Azmon, Moshava, Kibbutz and Moshav (Cornell Univ. Press, Ithaca, 1969).

10. A. Berler, New Town in Israel (Israel Universities Press, Jerusalem, 1970).

REGIONAL STRATEGIES AND THE EVOLUTION OF THE NEGEV URBAN SYSTEM

YEHUDA GRADUS and ELIAHU STERN

I. INTRODUCTION

Selecting a strategy of optimal spatial distribution of population in an unpopulated frontier desert environment is still a major theoretical issue in various parts of the world, but for Israeli regional planners it is an issue of real and immediate policy. It seems that in spite of the fact that each desert might be geographically and culturally a unique entity, some generalizations can be drawn from studying the Israeli Negev experience in the light of planning ideologies and the consequent regional strategies applied in the last thirty years for promoting growth and development.

It is particularly interesting to study the Negev at this time, shortly after the signing of the peace treaty between Egypt and Israel. The Negev desert will become the scene of rapid development in the coming years. Transferring Israel's army from Sinai will lead to a concentration of military installations and airports in the only large land reserve available for development. There is no doubt that the move will have a major impact on the industrial and urban system developed in the Negev. Before irreversible changes alter this system beyond recognition, it is worthwhile examining it in order to draw relevant conclusions for future development.

II. IDEOLOGY, PLANNING AND EARLY DEVELOPMENT

Ideology has had a substantial impact on the evolution of the urban and economic spatial system in the Negev desert. The agrarian socialist Zionist ideology adopted the so-called Central Place Theory[1] as a strategy for regional development to achieve the national objective of population dispersion.[2] This perhaps exemplifies one of the few cases where planning of an urban system was based on a geographical theory[3] which was perceived by the planners at that time as a socialistic approach for spatial planning. However, in such a new resource frontier area as the Negev desert where outflow from the region consists mainly of semi-processed raw materials, and inflow into the region consists of finished consumer or capital goods, the adoption of such a theory as a strategy for regional development indicates a lack of sufficient knowledge of the desert and its characteristics. The theory puts the emphasis on dispersal of population in hierarchical pattern, equity and balance regions; its applicability is limited mainly to the tertiary sector and restricted to consumer oriented activities. It assumes that input is available everywhere at the same cost. The pattern of input supply or input assembly is virtually excluded from consideration and for this reason there is no treatment of supply area. In cases such as the Negev desert, where the location of economic activity has been conditioned by localized raw material and a paucity of agricultural activities, the theory is of little practical value.

The planners realized that it would be difficult to establish towns in the desert solely as service centres, therefore they proposed that the towns would also provide housing for workers in the mining and other industries. Ten new towns were established within nine years in the arid Negev, based on the Central Place concept (Figure 1), with the regional first-order city, Beer Sheva, being planned to fulfill the functions of both a primary industrial and administrative centre.

By the early sixties it had become clear that there was a need for change, and that the hope held out for the development towns as service centres and residential areas for mineral workers had, for a number of reasons, not been realized. Expectations that the natural resources could form an economic basis were not fulfilled, and the number of workers in mining was too small to serve as a stable economic basis for the towns. As a result of water scarcity, agricultural development did not proceed; studies conducted to evaluate the success of townships concluded that they were not fulfilling their function as service centres and they were not interacting with their limited agricultural hinterland.[4] In certain cases, an ideological and social hostility developed towards the town whose population was small and characterized by instability and high turnover. The towns, which were populated mainly by new immigrants, quickly turned into pockets of poverty and the government was forced to come to their aid. It was obvious that a change was necessary in the planning conception and a new development strategy had to be adopted.

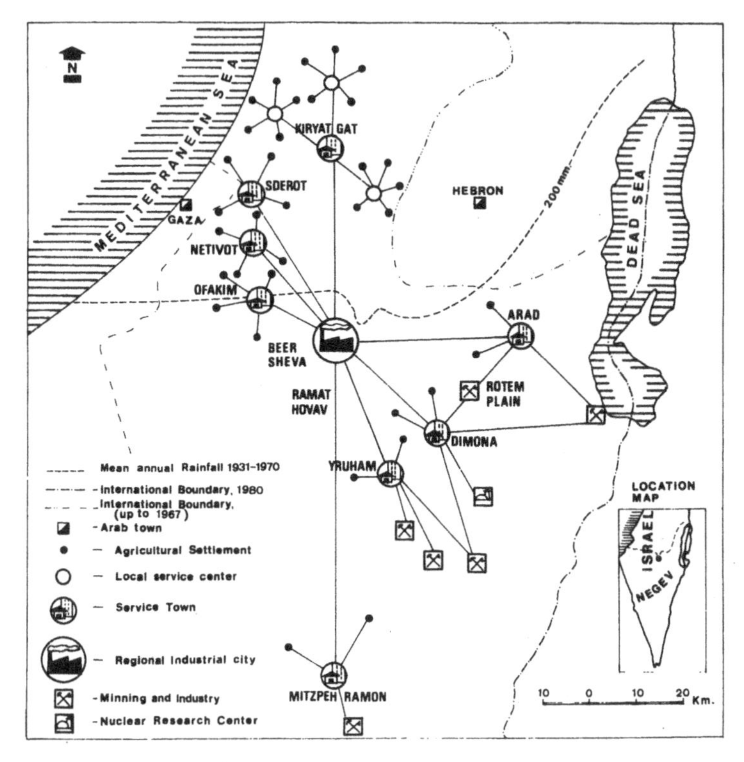

FIGURE 1, Regional Planning Concept of the Negev in its early stages. The concept was based mainly on the Central Place Theory.

III. EMERGENCE OF THE NEGEV GROWTH CENTRE REGION

Several actions taken by the govenment during the early sixties indicate in retrospect that the dispersed central place strategy has been officially abandoned. These permitted Beer Sheva to emerge as a growth centre,[5] a regional core, transmitting impulses of development to its periphery. Some empirical evidence indicates that "backwash" and "trickle down" forces[6] have been operating in the evolution of the system albeit in a sequential stage. The first action which indicates the change was the launching of a massive industiral-ization programme in the remote towns. Within a short time these towns which had been planned mainly as service centres became typical "company towns" based on footloose (mainly tex-tile) industries. This process created a great need for an urban economic infrastructure, i.e., business services, cultural centres and training institutions. Growth impulses from the region's industry were directed toward Beer Sheva and large job-providing organizations located their headquarters there. The spontaneous growth impulses (backwash forces) directed towards Beer Sheva were recognized by the government at a later stage and led to investment in one centre of promise and growth potential rather than in several areas of need. To this end, another three major governmental decisions were made. Beer Sheva was declared the captal of the southern district, providing jobs for thousands of employees in the tertiary sector. A new central hospital was dedicated for the entire region and the Ben-Gurion University was opened. The city was transformed into a service and commercial centre for the whole area, rather than evolving, as originally planned, into a central industrial city.

When the backwash force was operating, the relative weight of Beer Sheva in the Negev increased in terms of population from 39% in 1956 to 46% in 1965, following a population growth of over 155% (Table 1), and the city emerged as a self-sustaining growth centre. However, since the beginning of the seventies, the relative weight has decreased (43% in 1977), reflecting a growth of the periphery due to the trickle down force. It is clear from Table 1 that Beer Sheva's population growth rate during the period 1965-1977 was 54.9%, while the population growth rate of its periphery was 69.1%. In addition, the spreading of innovations relating to the development of human resources has developed between 1965 and 1977. For example, new branches of the University have opened in various towns along with the appearance of clinics as offshoots of the medical centre in Beer Sheva. The spread effect was accelerated by the government's dropping of incentives to Beer Sheva as a priority area. This resulted in certain enterprises relocating on the periphery. Thus, for the first time in the short history of the developing Negev, a massive migration from outside the region was observed, and even more important, the migration from Beer Sheva to its periphery was larger (although small in scale) than the reverse migration (Table 2).

TABLE 1, Population distribution for 1956-1977 in the Negev growth centre region (excluding Eilat and the Southern Arava Settlements

	1956		1965			1977		
	Population	Relative weight (%)	Population	Relative weight (%)	Growth rate 1956-65 (%)	Population	Relative weight (%)	Growth rate 1965-77 (%)
Total region	65,600	100.0	143.100	100.0	118.4	232,700	100.0	62.6
Growth Centre (Beer Sheva)	25,550	39.0	65.200	45.6	155.2	101,000	43.4	54.9
Periphery	40,050	61.0	77,900	54.4	94.5	131,700	56.6	69.1

Source: Population data compiled from the *Israel Statistical Yearbook*, 1957, 1966, 1978

TABLE 2, Total migration for the period 1971-1977 by origin and destination in the Negev Growth Centre Region (Negev GCR) (Key: *excluding Negev GCR, ** including unknown origins, IR irrelevant) Figures compiled from Ministry of the Interior - unpublished data.

from / to	Negev GCR	Growth Centre	Periphery	Israel*	Overseas**
Negev GCR	IR	IR	IR	74,646	20,161
Growth Centre (Beer Sheva)	IR	IR	3,987	29,311	8,667
Periphery	IR	4,135	IR	45,335	11,494
Israel*	53,848	20,459	33,389	IR	IR
Overseas**	1,787	414	1,373	IR	IR

IV. RECENT DEVELOPMENT - THE EVOLVING REGIOPOLIS

A new element of spatial order has come into being in the Negev which is not unlike Ghosh's Regiopolis concept for Indian Urban Centre Networks.[7] It has unified both centre and periphery within a single unit. The emergence of this phenomenon is a consequence of the spontaneous and directed forces which are currently operating, on the space economy of the Negev. It may be looked at as a third stage in the development process, after the backwash and spread effects.

It seems that the Negev Regiopolis tends to function as one metropolis. Instead of continuity of built up areas there are "islands" of small and medium-sized urban communities, and industrial complexes, separated by arid vacant land and connected by a network of roads. The intensive and independent commuting pattern (Figure 2) which has developed is an important factor determining the way in which

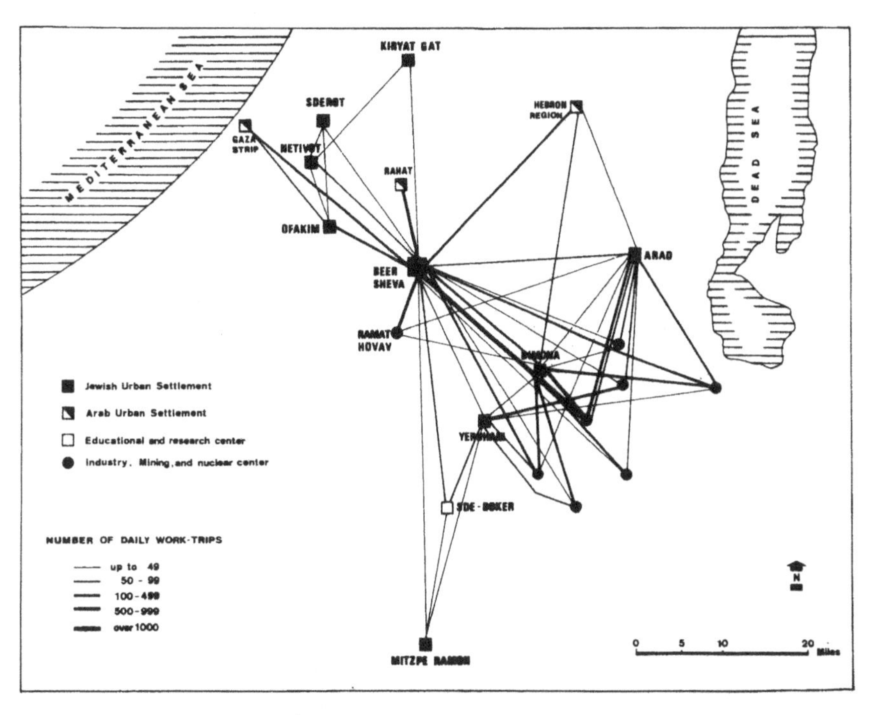

FIGURE 2, Commuting Pattern in the Negev, 1978

the Negev urban system functions as an integrated unit. Analysis of this commuting pattern indicates a clear radial pattern focusing on Beer Sheva. The main corridors of movement are from the towns to the industrial complexes and the research institutes in the south-eastern periphery, and from Arab areas in the north of Beer Sheva. To a lesser degree, commuting is also observable among the towns themselves. The pattern suggests that it is a mobile interconnected system of shared interests which acts as a single labour market area as well as a single service unit. The general decrease in the importance of distance in residential and place of work preferences[8] will further accelerate the emergence of this Regiopolis.

Based on the Negev experience, it is possible to define a regiopolis as an integrated set of cities, towns, villages and industrial complexes isolated in space from each other but interdependent and linked to a central city (not necessarily a metropolitan area) and among themselves, in such a way that any significant change in the economic activities, employment or population of one unit will directly or indirectly bring about some alteration in the others. It is a system of shared interests and a single growth area which taken together may be regarded as a growth centre.

The Negev Regiopolis (6.5% of the total population of Israel) offers a heterogeneous landscape consisting of a relatively big city, small towns, agricultural villages (of the Kibbutz and Moshav variety), industrial complexes, Bedouin settlements, recreation areas, research institutes and - exogenous to the urban network - military installations. All these are located in semi-arid and arid areas, no further than one and a half hour's drive from Beer Sheva.

V. FUTURE GOALS AND CONCLUSIONS

The existence of an integrated system of urban and economic activities (Regiopolis) should be taken into consideration in the future development of the Negev. Any significant change by the army will have an immediate impact on each subset in the system and on the system as a whole. Therefore, a comprehensive approach is needed and the appropriate unit for such a planning commission is the entire Regiopolis. The formulation of a regional development plan in joint consultation with all the relevant parties will be necessary to provide a common framework for decisions. Such a plan could promote the integration and the interdependencey needed in such a system. Planning is possible only by considering the system as a whole. This can be done only by a centralized institutional structure similar to the Tennessee Valley Authority (T.V.A.), The Corporacion Venezuela the Guayana (C.V.G) or the Superintendency for the Development of the Northeast (SUDENE) in Brazil.

Adopting the central place theory as a strategy for development in a desert environment indicates a lack of sufficient knowledge of the desert and its characteristics.[9] Our general conclusion from the Negev experience is that a compact and functionally interrelated system with a major centre is preferable in a desert environment. Israeli planners have generally found the growth centre strategy superior to that which requires equal spatial dispersion of investments in order to achieve economic equity.[10] The concentration of investments, especially in an unpopulated arid environment, in one selected centre will be likely to develop in the latter such advantages as economies of scale and this will benefit the entire region. High priority should therefore be given to the development of communication and transportation networks. Regional planners should be free from pre-conceived strategies developed in non-arid environments. Understanding the various elements of the desert is essential in applying spatial strategies.

VI. REFERENCES

1. W. Christaller, Die Zentralen Orte in Suddeutschland, translated by C.W. Baskin as Central Places in Southern Germany (Prentice Hall, New Jersey, 1966).

2. A. Sharon, (1951), Physical Planning in Israel, English summary, (Government of Israel Press, Jerusalem, 1951).

3. B.J.L. Berry and A. Pred. Central Place Studies: A Bibliography of Theory and Applications, (Regional Science Association, Philadelphia, 1964).

4. Y. Cohen, Urban Zones of Influence in the Souther Coastal Plain of Israel, (Settlement Study Centre, Publication on Problems of Regional Development, Rehovot, 1967).

5. F.D. Darwent, Environment and Planning, 1, 5 (1969)

6. L.G. Gaile, Geographical Analysis 11, 273 (1979).

7. S. Ghosh, Existics 267, 82 (1978).

8. L.P. Halverson, "The Critical Isochrome: An Alternative Definition", Proceedings of the Association of American Geographers, 7, 84 (1975).

9. Y. Gradus and E. Stern, "Changing Strategies of Development: Toward a Regiopolis in the Negev Desert", paper submitted to the Journal of the American Planning Association (1980).

10. M. Garon, City and Region, 4, 79 (1977).

EXTREME CONDITIONS AND RESPONSIVE ARCHITECTURE

ARIE RAHAMIMOFF

I. INTRODUCTION - EXTREME CONDITIONS IN ISRAEL'S DESERT

In a rapidly developing world, man's building activity is having an ever-increasing impact on the natural environment. Now more than ever before, there is a need to comprehend the environment as a total system. One such environment which is drawing increasing interest in an age of expanding populations is the deserts of the earth. There are many kinds of deserts, each with its distinct climatic and geographic features. Any attempts to settle these various kinds of desert will usually involve the creation of social conditions which have a bearing on the specific location and function of the settlement. Thus, architecture must be made to respond to this complexity of relationship between man and his varying kinds of desert environment. In the following pages we discuss a framework for such responsive architecture, choosing for illustrative purposes the rich variety of conditions that exist in the Negev and Sinai Deserts.

Israel's desert - the Negev - is small in size. Its surface area covers about 10,000 sq km. Yet, its peculiar location at the point of connection between Africa and Asia, its proximity to the Mediterranean and its definition by the Great Syrian-African rift create a variety of climatic conditions. Further extension of this variety is supplied by the neighboring Sinai peninsula, the Dead Sea and Jordan Valley and by the semi-arid zones on the northern edge of the Negev.

At some points the shift from one climatic zone to another is abrupt, while in other cases the transitional area may be quite broad. For example, the moderating effect of the Mediterranean on the coastal strip gradually decreases as we proceed towards the southeast into the desert. On the other hand, descent from the mountains of Sinai to the Red Sea coast is quite steep, and within a horizontal distance of 30 km. the difference in altitude can be over 2500 m. This results in major climatic differences between the regions of cold mountain desert and the hot coastal region adjacent to the sea.[1]

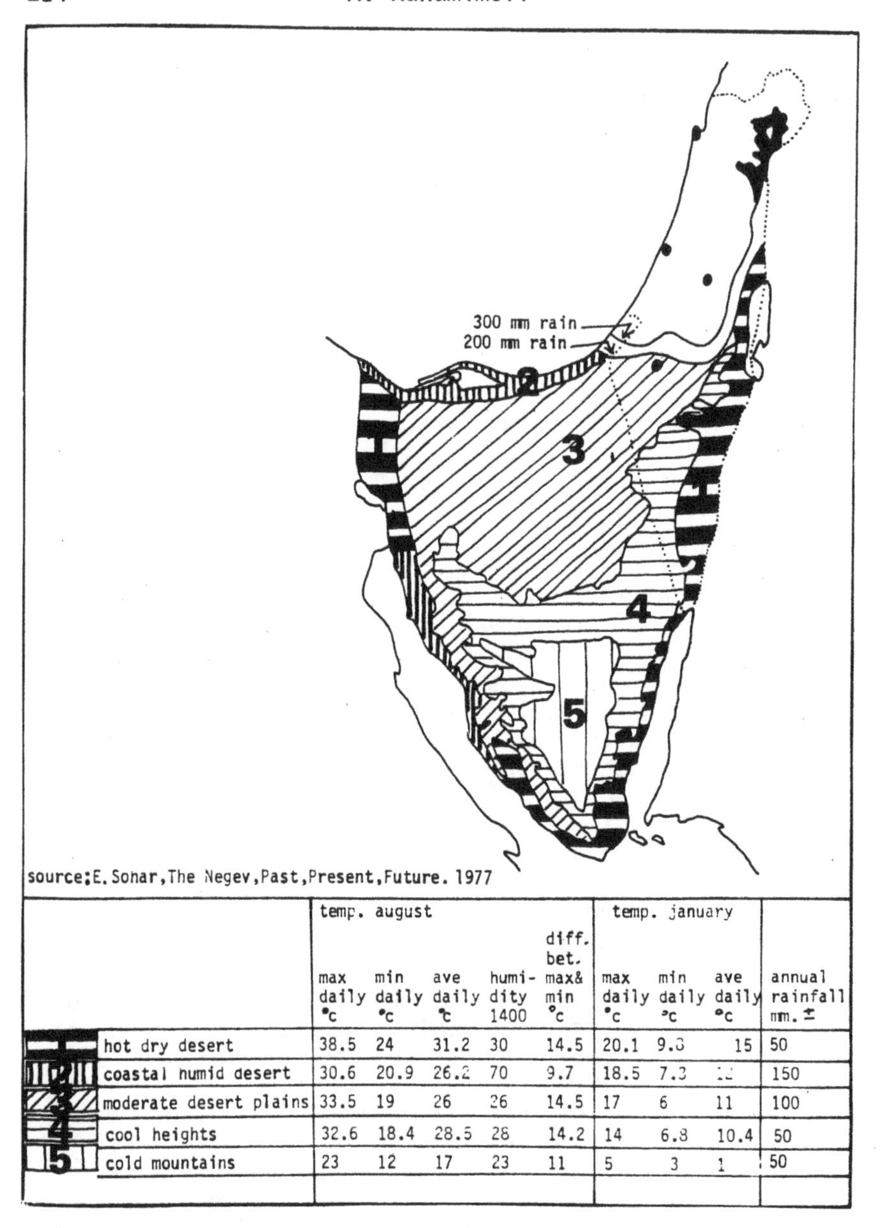

	temp. august					temp. january			
	max daily °c	min daily °c	ave daily °c	humi-dity 1400	diff. bet. max& min °c	max daily °c	min daily °c	ave daily °c	annual rainfall mm. ±
1 hot dry desert	38.5	24	31.2	30	14.5	20.1	9.3	15	50
2 coastal humid desert	30.6	20.9	26.2	70	9.7	18.5	7.3	[illegible]	150
3 moderate desert plains	33.5	19	26	26	14.5	17	6	11	100
4 cool heights	32.6	18.4	28.5	28	14.2	14	6.8	10.4	50
5 cold mountains	23	12	17	23	11	5	3	1	50

FIGURE 1, Classification of Desert Regions

According to one classification[2] there are five distinct desert regions in Israel and Sinai: (See Figure 1.)

1. HOT DRY DESERT: The Arava (that part of the Jordan Valley connecting the Dead Sea with the Red Sea) whose altitude varies from sea level to 400 m. below, and most of the Sinai peninsula's coast. Such desert is hot throughout the year, has very short and mild winters and very low humidity.
2. THE COASTAL TEMPERATE HUMID DESERT: A strip of land along the Mediterranean coast of Sinai not exceeding 10 km. in depth. Such desert has rainfall below 200mm/y. The relative humidity in summer is 70%.
3. THE MODERATE DESERT PLAINS: A wide zone between the coastal desert and the highlands. It reaches altitudes of 600 m above sea level. Rainfall is between 50-150 mm/y. Summers are generally mild with a relative humidity of 26%. During the winter heating is needed.
4. THE COOL HEIGHTS: The mountainous regions of the Negev and Sinai with altitudes from 600 m to 900 m above sea level. This desert has mild summers and chilly winters.
5. THE COLD MOUNTAINS: The high mountains of southern Sinai with altitudes from 900 m to 2600 m above sea level. This desert has cold winters and chilly summer nights.

As common denominators to all five of these desert regions one should notice:

a. Scarcity of rainfall;
b. High exposure to radiation;
c. Large diurnal and annual temperature fluctuations;
d. Sand and dust storm hazards.

As can be seen from the information shown above, in a rather small area one finds a variety of climatic conditions creating different types of deserts. But climate represents only one factor among many which establish the basis for regional development of the desert. The geographic features of the Negev create another set of constraints and opportunities.

The Arava valley with its warm winter climate together with its long stretches of flat surface make the development of irrigated or intensive winter agriculture a very attractive branch of economy.

The coastal strips and the mountain regions with their peculiar landscape features constitute important centres for tourism and recreation. Moreover, the relative proximity of the desert plains to the country's population centres creates a natural ground within the desert for the growing urbanization process of the country.

As already noted, the surface area of the desert is 10,000 sq. km and it covers about 50% of Israel. Yet this area is inhabited by only 7% of the country's population. Naturally the desert represents a reasonable goal for national population dispersion policy. However, it is precisely this point that creates such a challenge to the planner, namely: what should be the right type of development for the desert?

Obviously analysis of the social structure of desert population can help us to understand the alternative possibilities of planning the desert. As we have seen before there is a variety of extremes in climatic conditions as well as in geographic features. To this we should add the variety of social extremes.

In the Israel desert we can find pioneering kibbutzim and cooperative settlements dating back to the formative stages of the State. We can also find modern towns in which carefully articulated social structures, including an attempt at creating a proper physical environment and sound employment basis, ensures strong motivation of the population.

In the close vicinity of these two types of successful settlements, we find the less successful development towns. New immigrants were brought to these places without the proper preparation of absorption frameworks - neither physical nor social. The obvious result has been a largely transitional and alienated society. The proximity of the new and established towns, toegether with the well-to-do agricultural settlements to the culturally, economically and physically poor development towns stresses one more sector of extremities which is dominant in desert life in Israel.[3]

But it is the combination of this whole spectrum of extremeties - climatological, geographic and social as described above, together with technological, organizational and psychological constraints, which make the task of the desert planner such a complex challenge.

II. THREE DIFFERENT APPROACHES TO CREATING DESERT SETTLEMENTS

a. The Vernacular Approach to Extreme Conditions

Various desert cultures formed their settlements in an appropriate way to their environmental, climatological, geographic and social conditions.[4] In these settlements, which have evolved through a process of adaptation and development that has lasted for generations, man has succeeded in establishing architectural forms which articulate the various constraints and opportunities of the location. The proper functioning of these vernacular settlements is based on the interplay of appropriate physical frameworks and behaviour patterns of the society.

In North Africa, the Middle East, Central Asia and the south western United States, settlement types can be found which explicitly reflect their environmental stresses. The structure of these settlements, shaded streets and alleys, the clear definition of edges, enclosed squares, introverted dwelling types which are lit and ventilated from patios - all these are articulated in order to facilitate better living in extreme conditions.

Furthermore, the building components; thick walls, shaded patios and roofs, screened windows and a whole vocabulary of devices for natural ventilation are among the basic architectural features which serve to relieve the climatic loads on the desert settler.

In some remarkable cases, one can say that the climatic and environmental conditions, societal behaviour patterns and built form create an integral entity. Such exemplary achievements of desert architecture could have been accomplished only in societies where the whole community participated in the constant improvement of their environment. Through a process lasting many generations, the master builders evolved the professional expertise that would best serve the traditional and stable society.[5,6]

b. The Preconceived Approach to Desert Architecture

At the opposite extreme to the desert vernacular lies the architecture of the preconceived. Although there exists a few good examples of houses and developments adapted to local conditions, a distressingly large amount of present desert architecture in Israel bears the stamp of preconceived concepts. It should be remembered that both the planners and the inhabitants of the new desert settlements are newcomers to the desert environment, their first inclination was to look for and "transplant" planning schemes which are common in the milder climatic regions of the country.

A hierarchy of preconceived planning principles can be detected in:

i. THE SETTLEMENT STRUCTURE: Neighbourhood units are separated from each other with green belts that cannot be landscaped and maintained; or only at great cost because of the scarcity of rainfall and expense of water for gardening, low humidity and high radiation. The result is obvious; long distances of unprotected walks, exposure to glare and sand storms, expensive infrastructure with heavy maintenance costs. Such a settlement structure which is feasible in milder climates is simply out of place in the extreme conditions of the desert.

ii. ZONING:[7] The concept of zoning has a particularly large and negative impact on desert settlements. By spreading the basic investment too thinly over large areas, there is a reliance on vehicular movement between industrial, commercial, residential and recreational areas. The development of integrated urban quarters, in which a great part of the population lives, works and receives services, is particularly relevant to desert towns. Such quarters reduce the dependance of the population on traffic, thus decreasing the negative impact of the automobile on the settlement structure. It should be stressed that the introduction of traffic arteries and parking lots into the tissue of desert settlements reduces it from an environmental entitiy to insignificant fragments. The effect of self-shading of buildings is diminished, and the vastness of wide roads and parking lots expose the buildings to all the harshness of the desert climate. Later on we shall deal with the subject of appropriate movement patterns in desert towns.

iii. UNDEFINED EDGE CONDITIONS: The proper definition of the border between the built up area and the open desert is crucial to the successful functioning of the desert settlement. For a particular wind direction, orientation, slope and other site factors, a suitable edge condition should be designed to cope with the peripheral stresses on the settlement. Unlike the vernacular settlement, which devotes great attention to its edge conditions, in the preconceived schemes this consideration is usually neglected. Obviously, the lack of clear edge conditions has a negative effect on orientation, sense of place and the clear visuality of the settlement.

iv. HOUSING TYPES:[8] Apartment blocks on pilotis or pillars constitute the most common building type in Israel's desert towns. Imported to Israel in the nineteen-thirties to house new immigrants on the Mediterranean coast, it was further transferred to the various regions of the desert. The apartment block together with the extraverted individual house (another common building type in the desert) have far too great an exposure of their external surfaces to the unmodified thermal stresses of the climate. Only in a few cases where the need for ventilation is critical - e.g. the hot humid desert - is there justification for this fully extraverted type of dwelling. It should also be mentioned here that not only the housing types are "transplanted" from the north of the country, but so too are the building components - such as wall sections, window profiles, shading devices etc.

v. LANDSCAPING: The last point in this outline of preconceived architecture is related to the general treatment of public open spaces. The scarcity and high cost of water for gardening and the high evaporation due to the dry climate make the usage of intensive gardening prohibitively expensive. In most cases large open public spaces are simply left unmaintained. However, it has been observed that in those cases where small or moderately sized open spaces have been given to the inhabitants, they have generally been well kept. The combination of private open spaces, and small public open spaces with hard landscaping and shading, stands a much better chance of proper functioning in the desert than do large areas of public open spaces. Yet this type of low-rise medium density development is rather uncommon in the Israel desert.[9]

c. A Responsive Approach to Architecture

The vernacular and preconceived types of architecture represent the two extreme forms of creating a built environment in the desert. Naturally, between these two the ardent observer may find a whole set of intermediate approaches. These approaches occasionally provide local solutions to specific problems; or a "modified" vernacular in a contemporary context. However, the future prospects of development in the Israel desert embrace an objective of completely different scale, when seen in the light of the national policy to increase the proportion of Negev population from 7%

to 17% of the total population of the country, what this means is the addition of some half a million inhabitants to the present population of about a quarter of a million. And this national target is to be achieved within a period of less than one generation. Now even if this goal is accomplished at a slower pace, it will still require a major change in the existing pattern of settling the desert. Such development naturally, has enormous importance at the national and regional levels, but it also has unique architectural implications.

It is obvious that there is no ground or reason for the nostalgic search for a modern vernacular; there is no cultural or social framework and no technological or organizational basis for it. On the other hand, the application on a grand scale of preconceived architecture poses a dangerous alternative by ignoring regional and local factors and by attempting to standardize the built environment throughout the country. If this is done, it is likely that there will not be any environmental stimulus for new immigrants to move to the Negev. Israel should rather concentrate on encouraging a new type of regional development integrated with the specific conditions of the desert, and which will still fulfil the national goals of providing an arena for extended economic activity and supplying a residential base for larger population. Such considerations require a far more organized approach to planning; one which we may term Responsive Architecture; namely, responsive to climate, location and function.

III. RESPONSIVE ARCHITECTURE

Responsive architecure advocates:

THE REGIONAL	rather than	THE UNIVERSAL
THE LOCAL	rather than	THE IMPORTED
THE INDIVIDUAL	rather than	THE STANDARDIZED
THE EVOLVING	rather than	THE UNCHANGEABLE

In this section an attempt will be made to describe how different climatic zones, sites, technologies and societies may generate different settlement structures and built forms. This type of architectural approach conceives built form as the result of the various forces which are active in the process of creating a settlement, house, square or garden. Responsive architecture endeavours to establish a dialogue between environmental conditions, social characteristics and built forms. Thus, as environments change, the response of built form will change also and other built form responses may arise in similar environmental conditions as a result of different social structures of the inhabitants or the application of different technologies. The dynamic nature of responsive architecture allows for continuous improvement in living standards within the built environment, as well as in the incorporation of differences at the societal and personal levels. A further result of the proper balance between environmental conditions, and built form would be the establishment of an optimal framework for attempts to reduce the energy requirements of settlements both in the building process and in actual usage.

a. The Structural Heirarchy of Responsive Architecture

The relevance of responsive architecture may usefully be discussed in relation to "context" and "scale", for each of which there are three principle kinds.[10]

i. Three Contexts

A. The Climatological Context

If we analyse bioclimatic comfort charts of the different desert regions in Israel, as described previously, we find that each region should stimulate considerably different types of architecture. The hot humid desert presents climatic conditions that require proper ventilation during the hot periods of the year. This suggests the use of natural ventilation and careful definition of orientation, location and size of openings. However, sand storm problems may be ameliorated by a proper design of berms and vegetation.

In the moderate desert plains, winter cold may be a dominant feature to which the building and settlement will have to respond. Thus, proper location of the settlement on southeastern slopes is desirable. North to west winds may be disadvantageous during winter but desirable in the summer and therefore a scaled down almost continuous edge to these wind directions may give an optimal combination for winter and summer. On the other hand, the south east to south west orientation may have a soft, open and yet controlled edge that will allow low angle winter radiation to penetrate into the depth of the settlement. Houses may be densely clustered for low exposure during the summer. This will help to create shaded alleys in the public open spaces. The technology to be applied in the building should allow for high heat capacity in order to benefit from the diurnal temperature fluctuation both in summer and winter.

In the hot dry desert where summer winds may be too warm to reduce heat loads, settlement shading should be abundantly used in order to cool the air before it penetrates the houses. Proper selection and location of vegetation and evaporation pools may help to control the dryness of the air.

Indeed, the importance of appropriate vegetation can not be over emphasized. Trees that cast shade in the summer are very important for reducing the excessive radiation. Yet, it is only deciduous trees that will be appreciated in the close vicinity of houses during the chilly days of the winter. Vegetation can increase humidity in the air, filter sand storms and diminish the unpleasant effects of wind blasts. A proper functional integration of vegetation in the built environment can make a considerable contribution to the quality of life in the settlement.

Clearly such a brief description does not aim to give an exhaustive list of solutions for built form in different climatic zones. Its purpose is rather to illustrate the intricacy of climatic considerations that have to be interpreted in a properly responding environment.

B. The Technological and Organization Context

The different desert regions of Israel are characterized by an insufficient degree of technological development and infrastructure, the building industry and qualified manpower being mainly imported from the north. This factor presents a major difficulty in the development of local and responsive architecture, but one that may be tackled in stages:

i. Application of SELECTIVE RESPONSE - the building components may be of standardized type but their arrangement on the site, orientation and surface treatment made to respond to local conditions.

ii. Application of PROGRESSIVE RESPONSE - after the selective stage is accomplished and the basic technological infrastructure is established in the desert, improvements are inserted in the later stages of development while modification and adjustments are applied to the first selective stage.

iii. REVISION OF TECHNOLOGY - having in mind the magnitude and scope of settling Israel's desert, the existing building technology should be revised. The wide-spread usage of large prefabricated building components with whole rooms of complete wall panels, renders adaptation a difficult job. If smaller components were used, based on supporting structural and mechanical subsystems, adaptation and modifications would be much more feasible.[11]

iv. Application of RESPONSIVE STANDARDS - Israel building law sets national standards. The modification of these standards to local conditions would facilitate the creation of a more responsive environment. For example, in shading maps which were worked out for one development town, we found that at a critical hour in a summer afternoon, a certain neighbourhood had only 4% of its total surface area shaded. Furthermore, due to lack of proper design, the width of the shade at that critical hour never exceeded 1 m. and therefore even this amount of shade was left unused. Proper shading standards for different regions would establish a higher level of response.[12]

C. The Social and Personal Context

As already noted, the majority of inhabitants of Israel's deserts are immigrants from less extreme climatic regions. That will be also true for the future population of the desert. Individuals moving into new desert settlements, usually tend to look for physical environments, buildings, streets and gardens which are similar to the ones they have come from, but this creates conditions which are not climatically suitable and hardly maintainable in desert surroundings. As a typical example, one can mention the vast expanses of lawn in the desert kibbutzim. What is unthinkable to a vernacular society is considered desirable by an immigrant society. The "transplantation of life styles" and other corresponding building types is one facet of this attitude. Another facet is the illusion that if land is an ample commodity in the desert it can be parcelled and annexed to houses without major restrictions. This illusion is broken very soon after the house is inhabited, when glare, sand storms and high water bills start to enter. Only on rare occasions does the immigrant to the desert have the proper psychological preparation to bridge the environmental gap between his previous and future living conditions. If this observation is correct, desert architecture should facilitate a framework for a continuous social and personal response. Desert architecture will respond to the degree to which the desert is populated. With the growth and intensification of human settlements, factors such as isolation, poor amenities, lack of man-made objects in the visible environment and difficulty of perceiving scale in the landscape, will be diminished. This gives a framework for progressive, social and personal, response at all stages of the development of the settlement. An important point to note is that there will always be much uncertainty as to the future societal development of desert settlements. Thus it is imperative that desert architecture should at all scales be flexible enough to develop together with societal change. This refers to the whole settlement, the neighbourhood and the individual house.

ii. Three Scales

A. Response at a Regional Scale

The location of a settlement within a region is a response to various climatic factors that may be desirable or undesirable. Depending on the climatic region, decisions must be taken as to the desired penetration of winds, exposure to radiation and site topography. For example, if we compare a settlement site in the moderate desert plains to one in the hot dry desert; in the first case it is recommended that the settlement be located on the southeast slopes with medium wind penetration and a cross-sectional geometry that allows exposure to the winter sun and shading in the summer. On the other hand, in the hot dry desert valley, the preferred location would be on the northeast to east slopes, thus benefitting from minimum insolation at the hottest time of the day. The afternoon warm wind is allowed to penetrate the settlement after passing through shaded and planted areas, in order to cool it. It is obvious that local conditions can differ considerably at different sites within each broad region. Therefore careful climatological surveys should be carried out prior to design. In addition to climatological factors, attention should be paid to technological and sociological factors. For

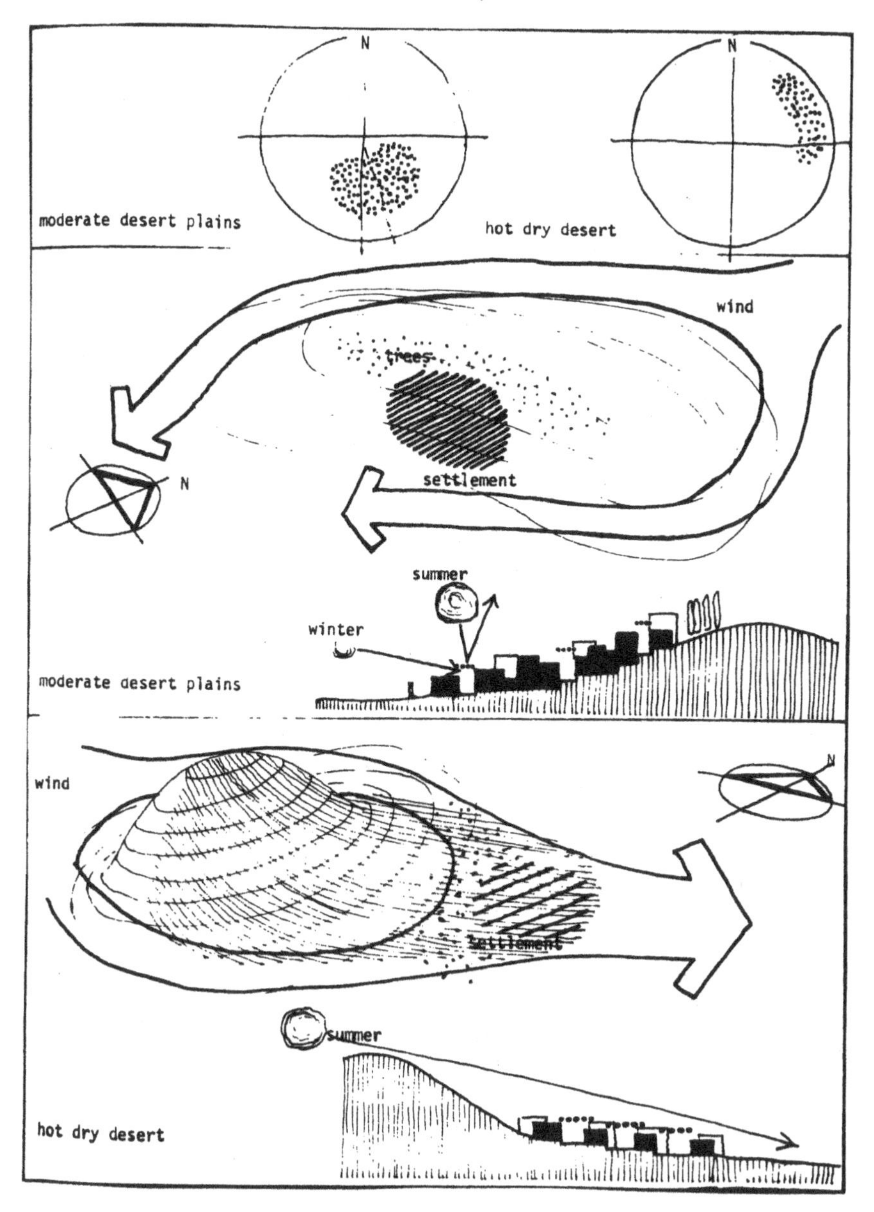

FIGURE 2, Region and Location

example, the location of the site within the existing road network might influence the building technology to be used.[13,14] (See figure 2.)

B. Response at the Settlement-Structure Scale

Settlement structure defines the interrelationship between built and void volumes, patterns of movement, edge conditions and settlement densities. In cases where it is necessary to reduce exposure to radiation and glare, and heat losses of the external envelope, shaded alleys may be created by forming the settlement structure as a continuous built tissue. This may be the case in the moderate plains. If, on the other hand, through ventilation is required, a looser settlement structure may occur, with less reliance on the self-shading of buildings and increased use of light shading devices. This may be the case in hot humid zones. The above two examples can be seen as the extremes of a whole spectrum of solutions which respond to a variety of climatic conditions.(See figures 3, 4.)

Then of course, superimposed on these considerations must come various sociological factors. For example the kibbutz, a communal settlement as compared to a conventional agricultural settlement, would require quite different approaches to definition and use of private spaces, and hence settlement structure. A well balanced settlement structure would take advantage of a hybrid composition of different basic structures which react to local conditions.

C. Response at the Basic Unit and Cluster Scale

Each individual basic unit should be considered together with the development possibilities of its immediate surroundings. The combination of these has a certain set of properties in terms of:

i. Growth to meet the needs of different and changing family sizes and life styles, as well as different approaches to building, such as self-help or cooperative construction.

ii. Hierarchy of private and public spaces.

iii. Size and intensity of treatment of open spaces.

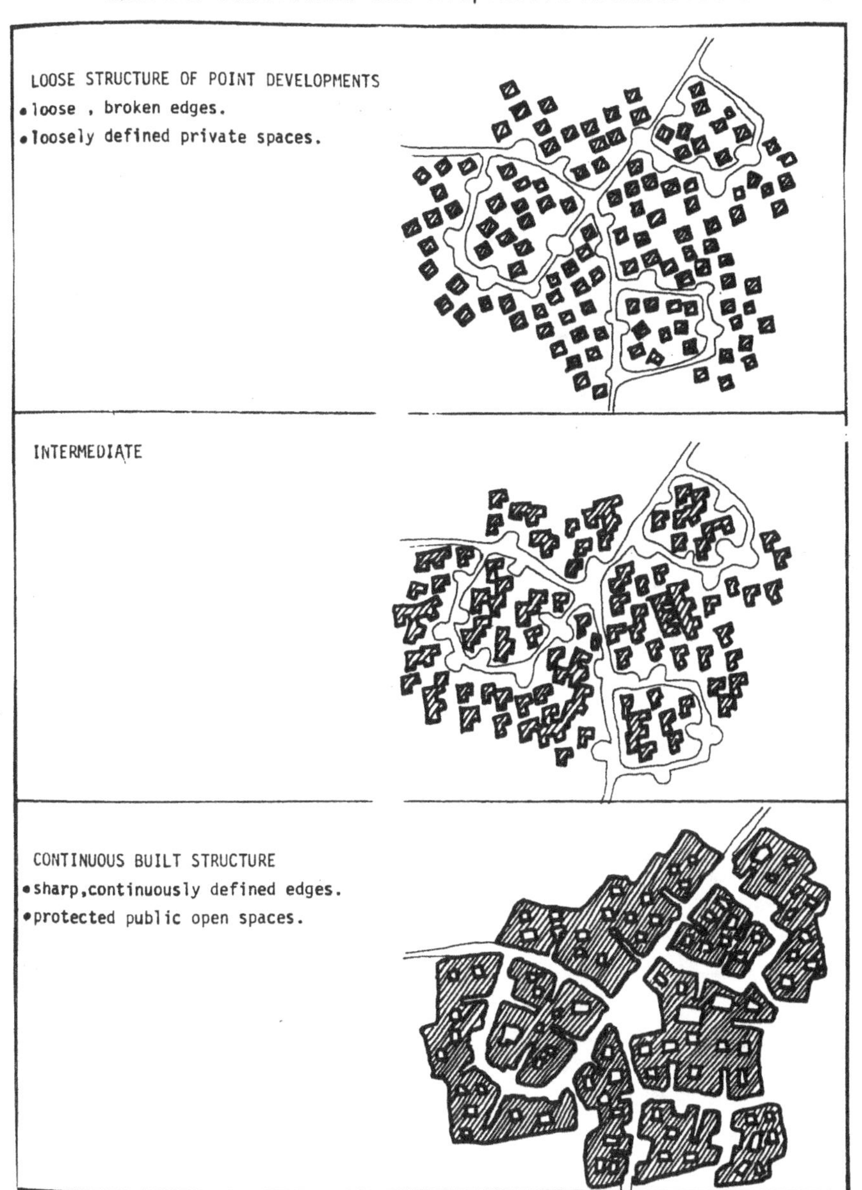

FIGURE 3, Settlement Structure

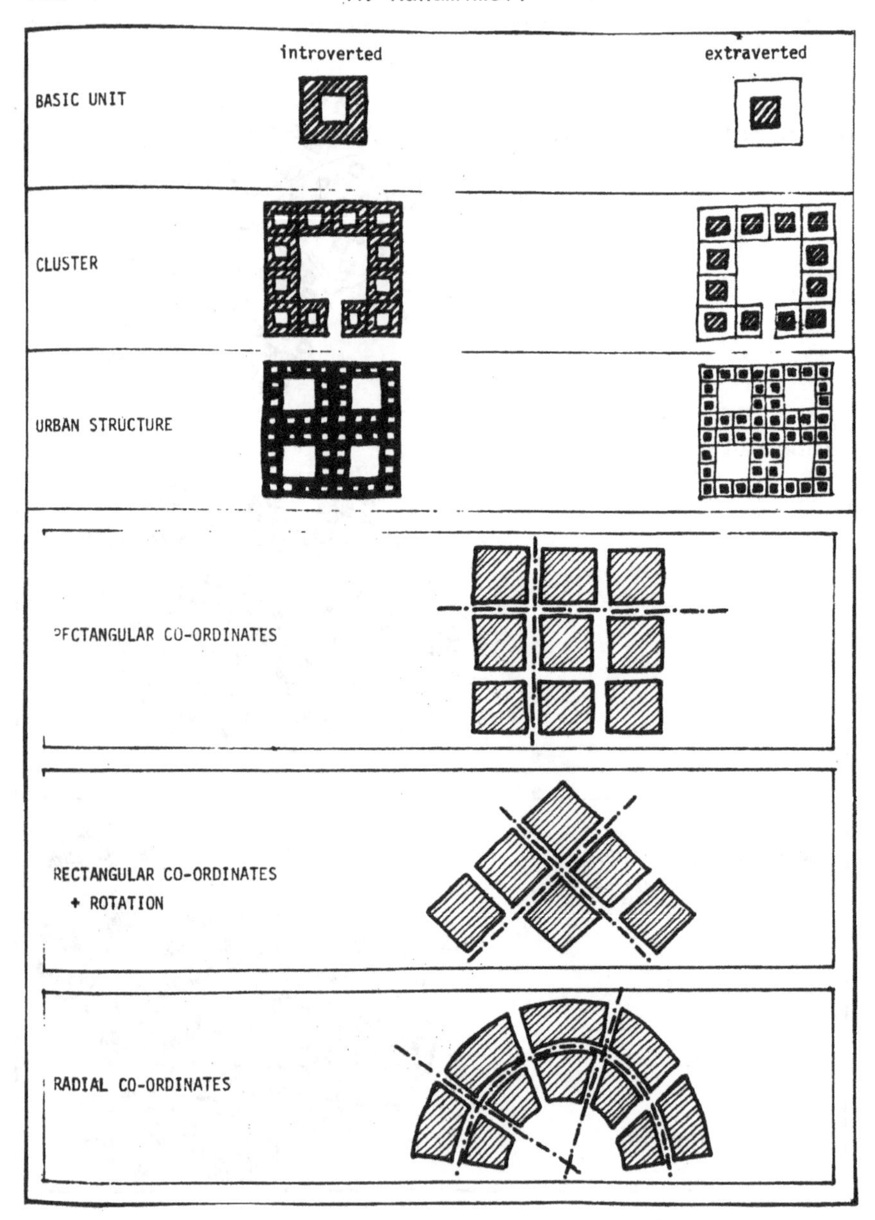

FIGURE 4, Morphology of Urban Structures

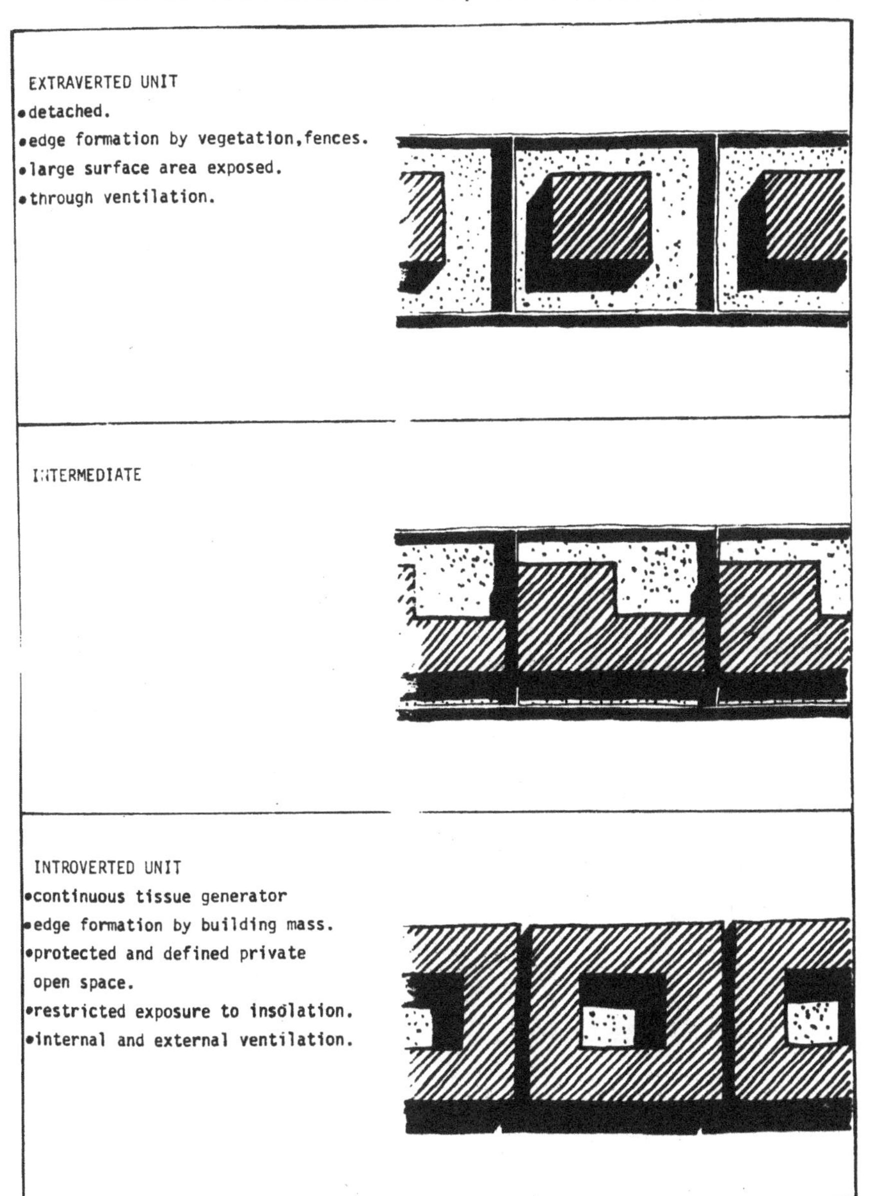

FIGURE 5, Basic Unit

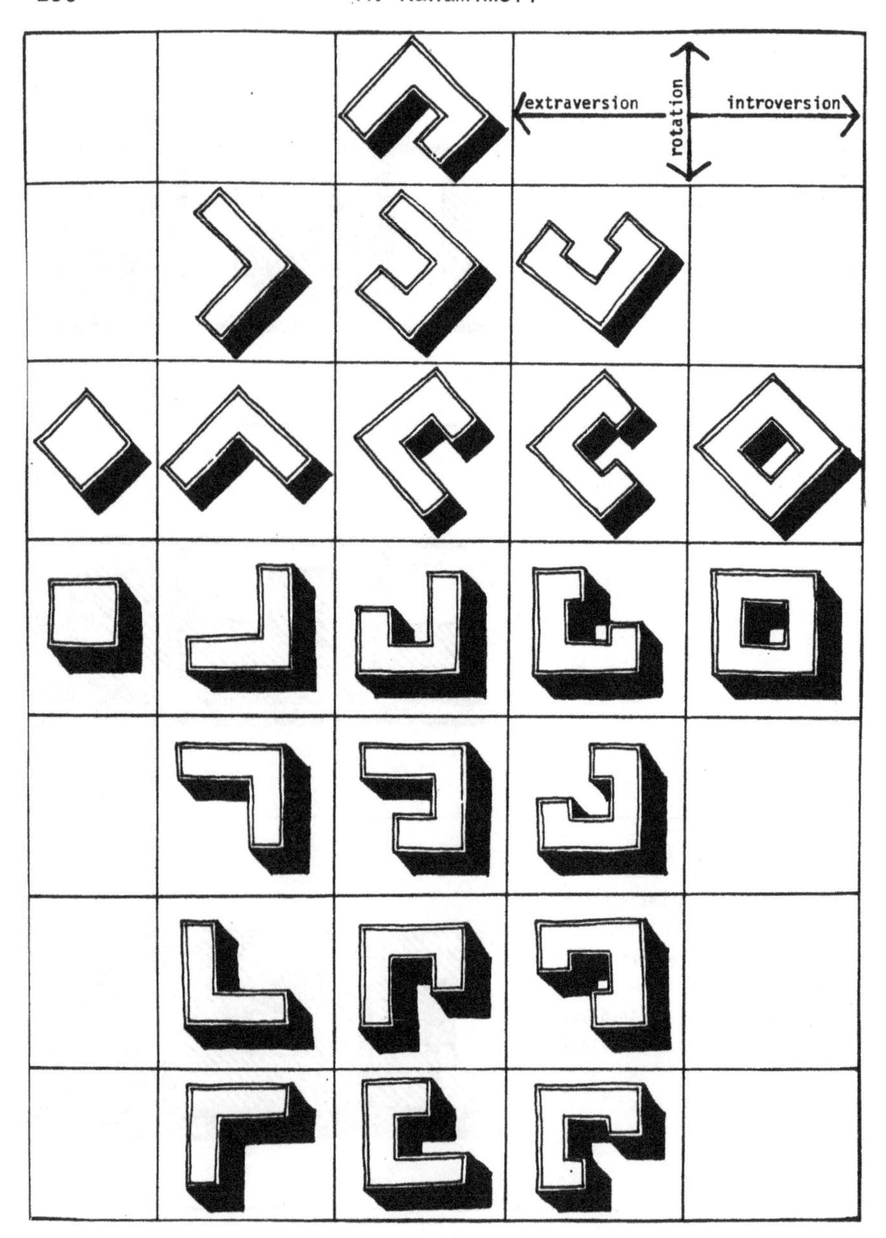

FIGURE 6, Transformations of Basic Unit

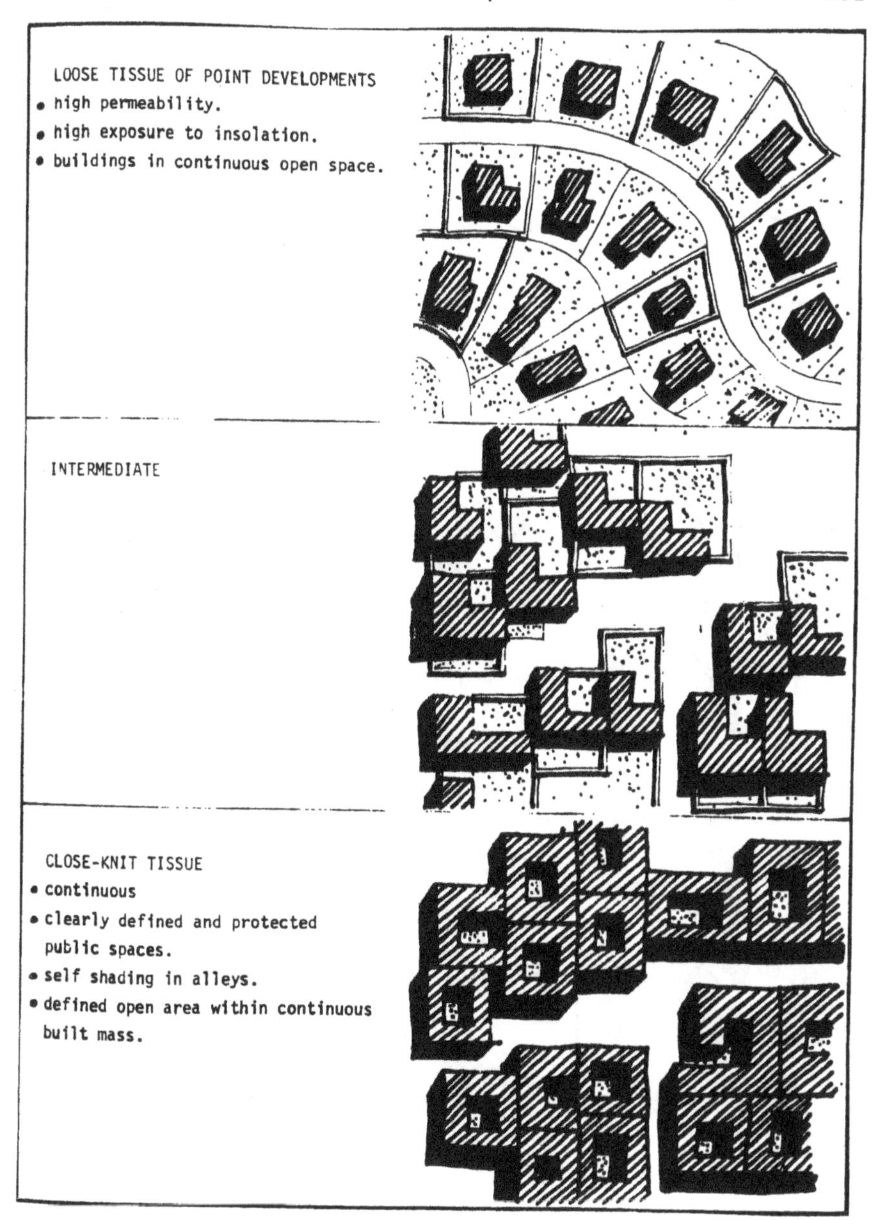

FIGURE 7, Cluster

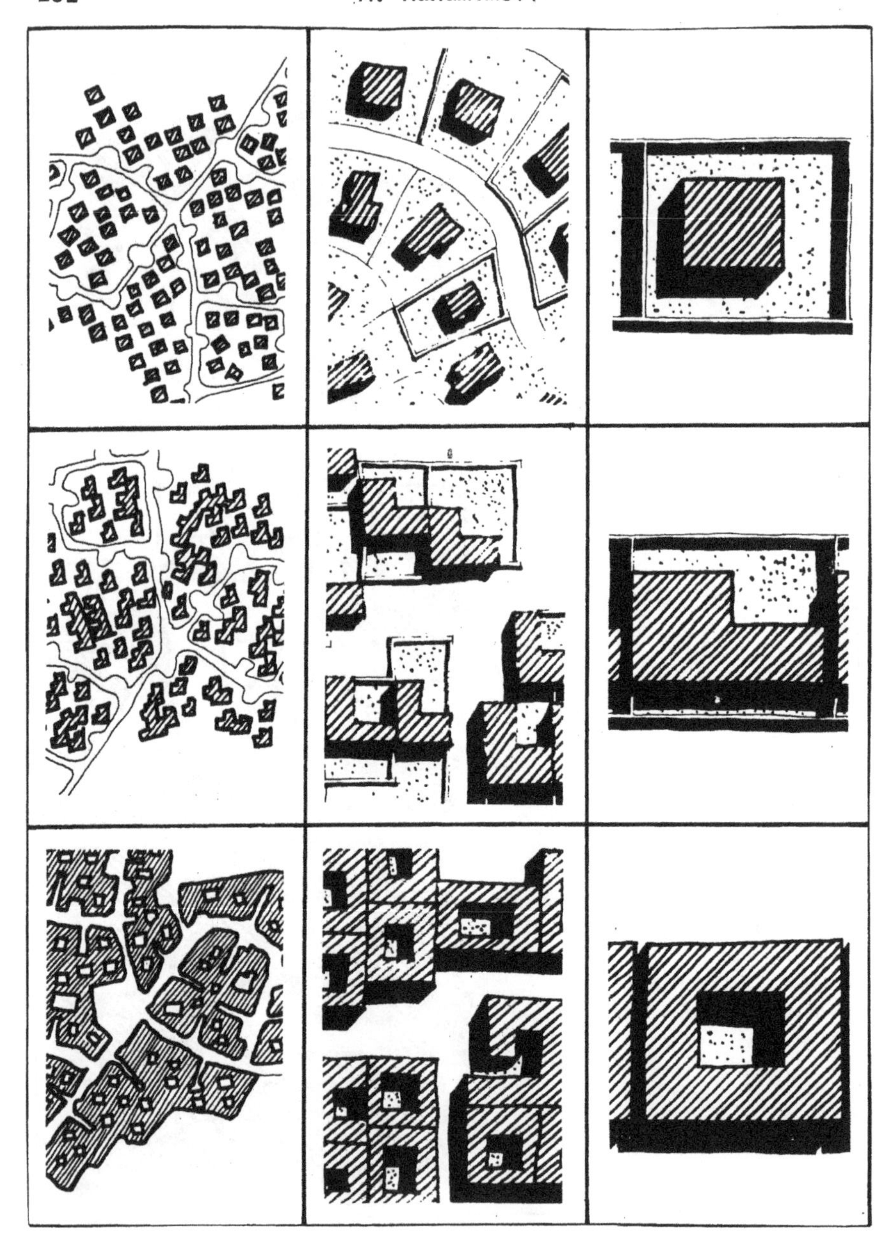

FIGURE 8, Settlement structures as Clusters of Basic Units.

compact geometry of traditional urban pattern in arid zones, shaded alleys and squares.

contemporary lack of shading concepts and norms. superfulous unprotected open public spaces.

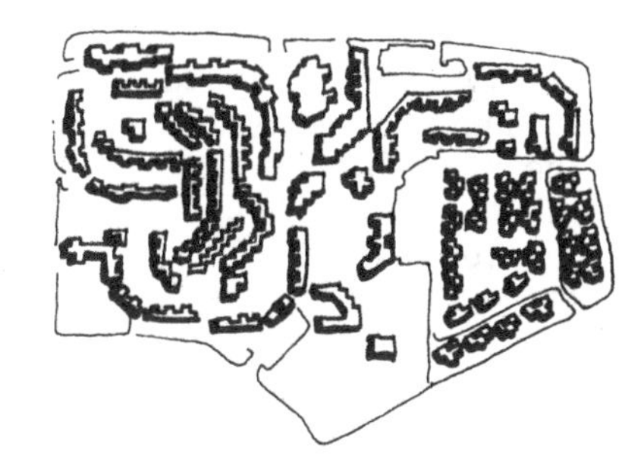

supports for public shading devices

FIGURE 9, Shading Patterns

iv. The degree of extra or intro-version responding to the relative extremeness of the climate. By different treatment of parcelation edges; a basically extraverted unit can be given an introverted quality. Vegetation, fences and walls constitute a whole range of edge-definers with different permeabilities.

The cluster scale allows the development of microclimates in public spaces. Clear concepts of public shading, integration of carefully specified vegetation to suit climatic conditions and movement patterns within clusters, facilitate the definition of dimensions and proportions of the public spaces. A great deal of literature and experimentation exists on the subject of climatic adaptation of building components, yet in Israel, little of this has been implemented. An important issue in this field is the integration of passive energy systems into housing. Various methods are at an experimental stage and there is no doubt that in the near future some of them will play an integral role in responsive architecture. A further issue is the development of a vocabulary of building components appropriate to the desert, such as window profiles which prevent dust penetration. (See figures 5,6,7,8,9.)

b. Responsiveness Related to the Definition of the Edge Condition

Because of the environmental extremes of the desert, the design of appropriate edge conditions of settlements, clusters and basic units should be an important part of the overall design. The edge is the border zone between the IN and OUT. It is the ultimate expression of the response of the built form to external environmental factors while it simultaneously reflects internal forces. Strong, disruptive winds,frequent sand storms or excessive radiation would encourage low perforation of windows in the external wall. However, a less extreme external environment might stimulate the usage of large openings; and this may be further encouraged if the dwelling faces a particularly beautiful landscape (an example of an internal force). If climatological factors set a demand for high heat capacity of the envelope, thick and massive exterior walls might be the appropriate response. (See figures 10,11.) A similar set of considerations will be taken into account when the settlement edge is to be defined. A parallel analysis of environmental stresses together with an awareness of behaviour patterns of the inhabitants may result in the definition of a hard border zone between the settlement and

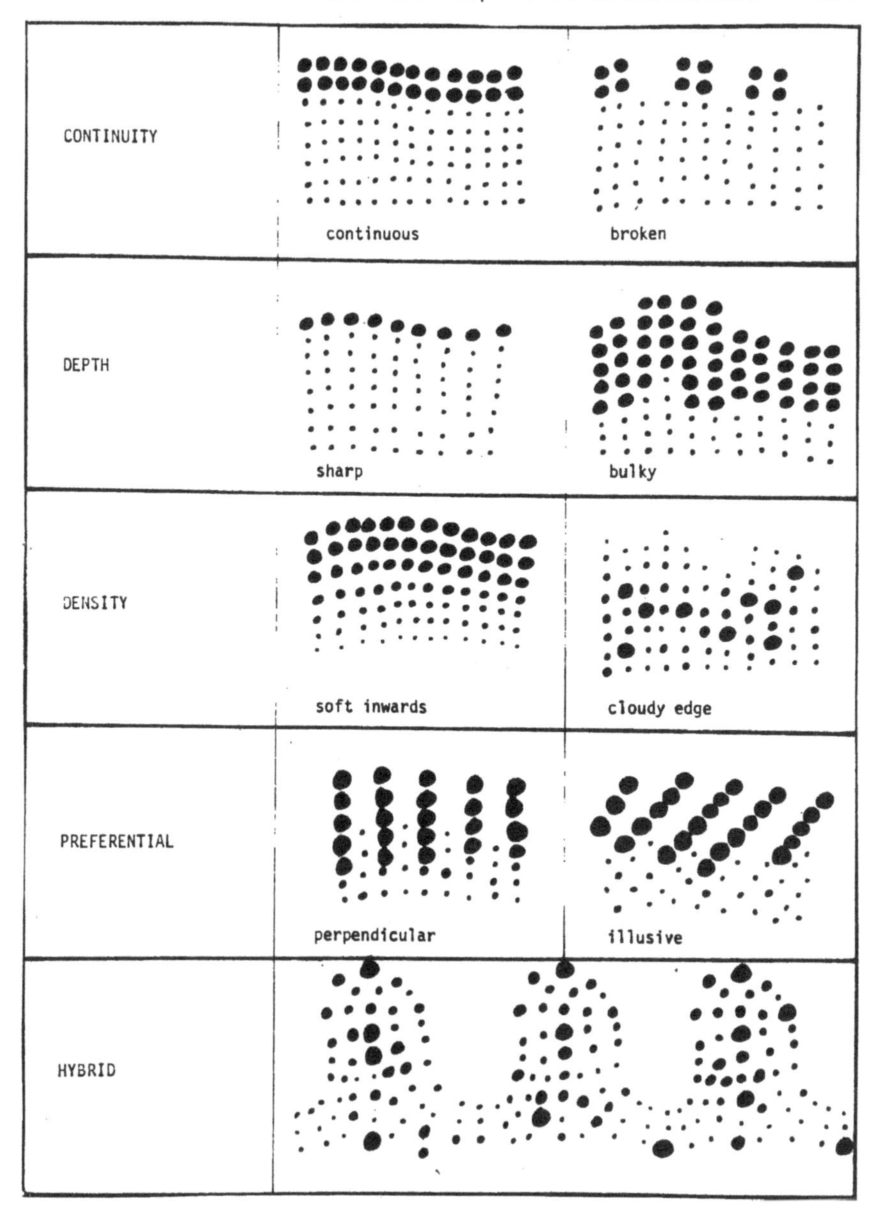

FIGURE 10, Morphology of Edges

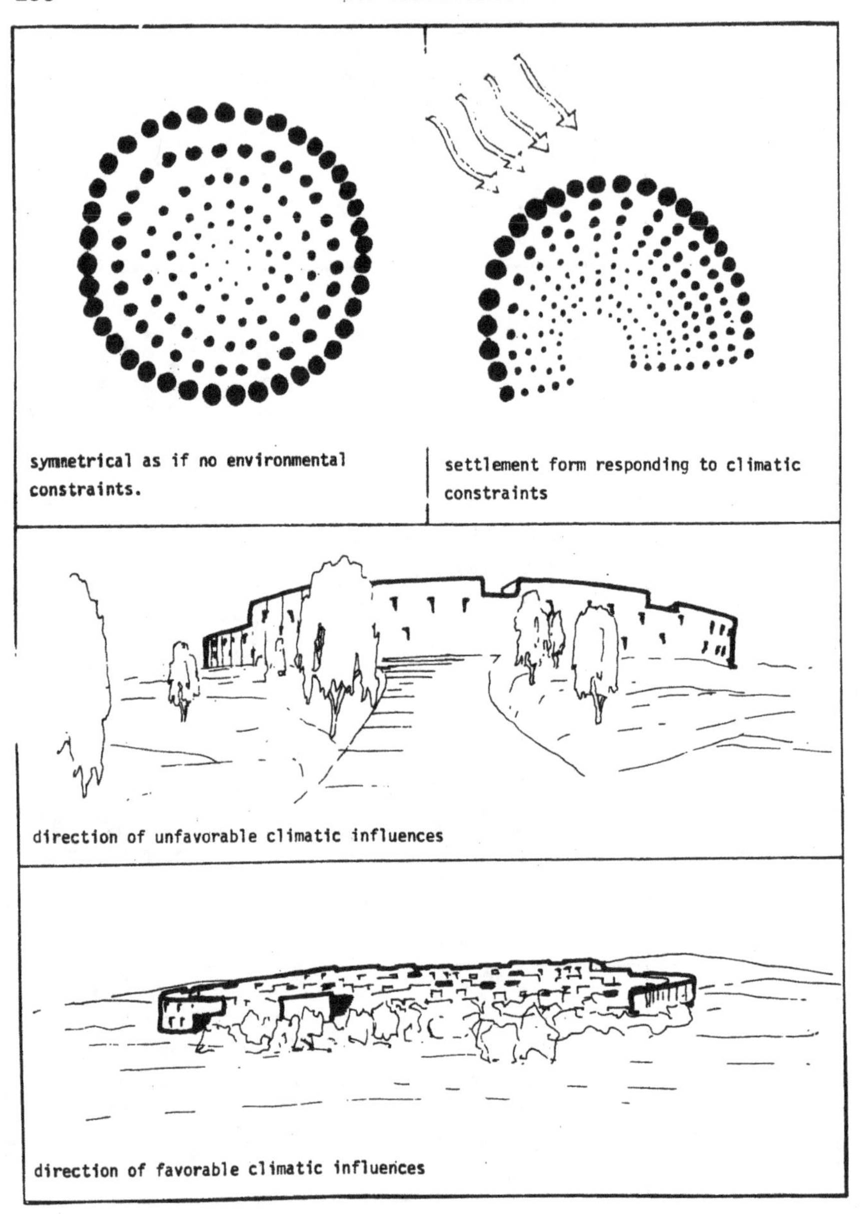

FIGURE 11, Formation of Edge, Image

the desert. Settlement edges may be continuous or broken, thin or bulky, soft or hard. Edges can also be hybrid. Various building and landscape features can create the edge. Buildings can be high or low, point blocks or row houses, continuous apartment blocks or individual villas. The factors that define the type of house to be used are not only programmatic, internal or those resulting from land use policy. They are also a reflection of the external environmental stresses. Edges can be defined by carefully selected vegetation that will reduce glare, deflect wind blasts or give visual definitions to the site. Berms, terraces and earthwork can also contribute to specific demands from edge conditions. The responsive use of appropriate edges for different orientations, sites and functions, will give each site its singularity and by doing so will contribute to the growth of local architecture.

c. Framework for Responsiveness

The issues discussed above cover a broad range of scales and contexts and they relate to many aspects of desert life. In order to integrate these issues in the continuous process of implementation of architecture in the desert a new framework for responsiveness has to be created. Its aim should be to overcome the present estrangement between "producers" and "users" of built form which exists in many parts of the world. This is especially deliterious in the desert, since the producers are located far from the users and in different conditions. In such a context it is especially critical to co-ordinate; decision makers, planners, architects, promoters and contractors, social workers and administrators together with a range of other professionals. One must of course, not neglect the continuous and multi-faceted co-operation of the present and future inhabitants of the desert. A continuous dialogue among the various participants in this grand-scale endeavour of settling the desert is essential if success is to be achieved in creating a suitable environment under extreme conditions. Modern society has lost the capacity of developing its settlements on the trial and error basis characteristic of the vernacular societies. At the same time, great quantities of built form are accummulating for the future. A bold change in attitudes - from the preconceived to the responsive - will transform the modern desert society into a group of people participating in an exciting experiment of constant dialogue with their environment. The architect in this situation has a much more responsive role than traditionally. He has to be the interpreter of extreme environmental conditions into built form, and must also design the physical settlement structure to reflect the expectations of different immigrant groups and individuals. In order to achieve this goal, the architect must transform himself from being the remote professional and become the local expert. This should be simultaneous with the creation of locally based planning teams from all relevant disciplines. Furthermore, the physical implementation of designs should be carried out by building groups who are thoroughly familiar with the requirements of desert settlements. Appropriate education and public consciousness towards desert living should also be instituted at all levels. If the above concepts can be realized we may stand a chance of greatly upgrading the quality of desert living.

IV. ACKNOWLEDGEMENT

This paper is based in part on a lecture delivered while the author was a guest of the Technische Universitat, Berlin. He wishes to thank: J. Berkowitz, G. Sifrin and P. Kiczales for their assistance in preparing the manuscript.

V. REFERENCES

1. M. Evenari, L. Shanan and N. Tadmor, The Negev: The Challenge of a Desert. (Harvard Univ. Press, Cambridge, Mass. 1971).

2. E. Sohar in The Desert: Past; Present and Future. ed. E. Sohar, in Hebrew (Reshafim Publications, Tel Aviv 1977).

3. E. Spiegel, Neue Stadte in Israel/ New Towns in Israel, German and English (Kramer Verlag, Stuttgart 1966).

4. B. Rudofsky, Architecture without Architects, (Doubleday, New York, 1964).

5. S.H. Al-Azzawi in Shelter and Society, ed. P. Oliver, (Barrie and Rockliff: The Cresset Press, London, 1969).

6. A. Rapoport, House Form and Culture (Prentice Hall, Englewood Cliffs, 1969).

7. P. Blake, Form Follows Fiasco (Little Brow, Boston 1974) A. Rahamimoff, A. Halevy, D. Guggenheim, A. Abrahami, A. Block, I. Kleper, N. Gviniashvilli and S. Rahamimoff. J. of the Ass. of Engineers and Architects 7-8, 48 (1978).

8. A. Rahamimoff, I. Haissman and S. Rahamimoff, Architectural Design XLV-7, 438 (1975)

9. G.O. Robinette, Plants, People and Environmental Quality (U.S. Dept. of the Interior, Natl. Parks Service, Washington D.C. 1972).

10. See for example: L.L. Whyte in Structure in Art and Science ed. G. Kepes (Studio Vista, London, 1965). A. Koestler, The Ghost in the Machine (Pan Books, London, 1967)

11. R. Bender, ITCC Review IV-1, 42 (1975).

12. N. Bornstein, Yeruham: Study of Desert Urban Fabric, (Institute for Desert Research, Sede Boqer 1977) unpublished.

13. V. Olgyay and A. Olgyay, Design with Climate (Princeton Univ. Press, Princeton, 1963).

14. K. Lynch, Site Planning (MIT Press, Cambridge, Mass, 1971).

HEATING AND COOLING VIA THE UTILIZATION OF NATURAL ENERGIES

BARUCH GIVONI

I. INTRODUCTION

In approaching the problem of heating and cooling desert buildings, two factors have to be considered. One is the desire to protect the inhabitants from those climatic extremes which are inherent to desert regions. The other is the importance of keeping the heating and cooling equipment as unsophisticated as possible. For in general the remoteness of desert areas will imply increased equipment cost owing to transportation charges and corresponding difficulties in maintenance and repair.

Fortunately it is possible to keep sophistication to a minimum if one makes maximal use of the natural energies that desert climates afford both for heating and cooling purposes. Our approach is one of so-called integrated or passive heating and cooling systems. This method uses the various building elements such as windows, roof, wall, and even the ground under the house as actual elements for heating and cooling. In this way the building itself becomes an "energy machine" and the architectural and structural details become integral parts of the components of the energy system.

II. CLASSIFICATION OF PASSIVE HEATING SYSTEMS

Various types of passive solar systems have been developed recently by several investigators.[1] In this review some of them will be described and some of the problems encountered in their application will be discussed.

The term "passive", in the framework of the present section does not exclude the use of mechanical equipment such as fans, pumps, etc., for the transport of the energy. Rather it emphasizes the application of such equipment when it seems desirable but as aid to the natural "passive" processes of heat transfer and in the context of a "soft technology" approach.

It is possible to classify the different passive energy systems for heating of buildings into the following main classes:

a. Direct heating of the inhabited space by penetrating solar radiation.
b. Utilization of sun-porches and attached greenhouses as solar collectors.
c. Utilization of southern walls as solar collectors.
d. Utilization of the roof as a solar collector

Such classification has a special advantage from the point of view of choosing the appropriate system and adjusting special design details of the energy system to different building types. This point is especially important in integrated systems which use 'ordinary' building elements as integral components of the energy system.

Additional advantage of such a classification is that it clarifies the options for combinations, within a given building, of different <u>solar heating</u> systems as well as combinations of different <u>heating and cooling</u> systems.

Each one of the above main classes can be sub-divided into many different specific systems. Some of the systems already in use are the following:

<u>Direct heating by penetrating solar radiation</u>

- Southern windows as solar collectors (and eastern/western walls with special design details)
- Roof skylights with southern orientation

<u>Sun-porches and attached greenhouses</u>

- Southern sun-porches (solariums)
- Attached greenhouses

<u>Southern walls serving as solar collectors</u>

- Masonary walls acting as combined collection/storage systems (Trombe walls).
- Insulated lightweight walls acting as solar collectors, e.g., containerized water, behind a southern glass wall or under roof skylights, acting as combined collection/storage system (e.g. water-filled barrels, cylinders, etc.)

<u>Roof as energy collectors</u>

- Lightweight inclined roof panels serving as solar collectors.
- Water bags with movable insulation over horizontal metal roof (Skytherm).
- Roof Radiation Trap (Givoni)

It is possible to combine, within any single building, several of these solar heating systems. Furthermore, they can be combined with different natural cooling systems such as night cooling of the building mass, convective and evaporative cooling of a gravel storage, nocturnal longwave radiant cooling, etc., as will be discussed below.

a. Direct Heating by Penetrating Solar Radiation

Direct heating of the inhabited space by penetrating solar radiation was the first approach applied in the U.S.A. in the so-called 'solar houses' of the thirties.[2] In these houses, large glazed areas were exposed to the south, admitting solar energy directly into the buildings. However, the performance of this 'first generation' of solar buildings was disappointing. On sunny days in winter they overheated even in cold climates and after sunset they quickly cooled down, requiring full-capacity conventional heating during the night.

Two main reasons were responsible for this poor performance. The first was the lack of adequate heat capacity in the typical stud-wall lightweight American houses. The second was the high rate of heat loss through the large glazed areas during the evenings and nights. In addition, rooms without direct southern exposure were too cold and often required heating on cold days even during sunny hours.

The lesson learned from the analysis of the thermal performance of these 'solar houses' with respect to the design of direct penetration solar heating systems was that during the sunny hours, southern windows and glass walls can act as very efficient solar collecting systems, provided the following conditions are met:[3]

i. Sufficient heat capacity is necessary in the building to absorb excess solar heat during the sunny hours, releasing it in the evening.
ii. Means enabling a great reduction in rate of heat loss through the large southern windows during the evenings and night hours are essential.
iii. Means are necessary to transport excess heat during the sunny hours from the sourthern to northern rooms in the house.

One of the problems encountered in buildings heated by direct penetrating solar radiation is the fading of fabrics, wallpapers, furniture etc. by the high level of sunlight.

FIGURE 1, D. Wright house in Santa Fe. N.M. Direct gain through southern glass wall. Adobe walls for thermal storage. Folding insulating panels. Supplies 90% of heating needs.

The high efficiency of southern windows as solar collectors results from the fact that the collecting efficiency increases as the temperature of collection decreases. The temperatures at which solar energy penetrating through windows is collected is the lowest possible, namely at the human comfort range.

Therefore, southern windows designed as solar collectors can be considered as the least expensive and most efficient, particularly in view of their other contributions to the quality of the indoor environment in terms of contact with the outdoors, natural light and view, as well as to natural ventilation in summer. However, the low temperature of collection makes the problem of thermal storage more complex and expensive to solve.

In addition, when southern windows are used as solar collectors there is the problem of transporting heat from southern to northern rooms. This heat transport can in some cases be effected by natural convection, but in most cases it requires the use of a small fan (the air-flow resistance in this case can be very small).

To enable effective utilization of the solar energy penetrating during the daytime hours through southern windows it is necessary to absorb the excess heat in a way which will prevent overheating of the indoor space during daytime and which will release the absorbed heat later at night. In practice it is possible to store the excess heat during several sunny days in such a way that it would be possible to heat the dwelling not only during the night of the sunny days but also during one or two subsequent cloudy days.

It is possible to absorb and store the excess heat from the penetrating solar radiation in two ways. The first is essentially a passive one, namely within the structural mass of the building: in the floor, ceiling, walls, partitions etc. The path of heat flow into the mass is mostly by natural convection and radiation, both in the form of sunlight reflected from the irradiated surfaces to other surfaces and in the form of longwave radiation emitted by the warmer to the colder surfaces. The stored heat is released back into the indoor air at night also in a passive way and reduces the rate of cooling of the dwelling.

The second way to absorb and store the excess solar heat is by blowing the warm air through a thermal storage of gravel or small containers of water. This is an active form of heat transfer. The gravel serves both as a thermal store and as a heat exchanger.

The heat stored in the gravel can be released when needed by blowing the indoor air in a closed loop through the warm gravel. Thus, the active way of storage enables better control and more efficient utilization of the stored solar energy, but requires a mechanical system (at least a fan) for its operation and may be more expensive than the passive way.

b. Sun-Porches and Attached Greenhouses

It is possible to design buildings with balconies covering all or most of their southern facade. If such balconies are tightly glazed in winter and ventilated in summer (by opening or removing the glazing) then the same balcony serves as sun-porch (solarium) in winter and as an overhead shading of the southern wall in summer. During the winter season, solar radiation is trapped in the sun-porch and can even penetrate into the building if so desired, through suitable doors and windows thus heating the rooms adjoining the sun-porch.

FIGURE 2, D. Balcomb house in Santa Fe, N.M. Attached greenhouse as solar collector. Adobe wall for passive thermal storage and a rock bin under the floor for active storage. Supplies 84% of heating requirements.

The wall between the sun-porch and the rooms can itself be built as a solar collector, either as a masonary "Trombe" wall (see later), as an insulated lightweight collector-wall or as an ordinary insulated wall.

The warm air from the sun-porch can flow passively into the adjoining rooms, when the doors connecting them are open, or can be blown by a fan to northern rooms, or any other room not connected directly with the sun-porch. It is also possible to circulate the warm air through a gravel storage within the building or attached to it, and to use the stored heat for heating the building at night. The transport of the heat from the sun-porch into the building or the thermal storage also helps in preventing too-high temperatures in the sun-porch. Such ventilation can be activated thermostatically whenever the sun-porch temperature exceeds, e.g. 25°C or any other desired level.

If it is desired to use the sun-porch as a living space in the evenings during the winter, it has to be equipped with operable insulation, which is open during the day and closed during the evenings and nights so that the porch temperature is kept close to the indoor temperature.

One of the main advantages of sun-porches over other solar collecting systems is that the 'collector' serves also as a useful living area, both in summer and winter. It can serve also as a greenhouse for growing various plants whether in winter or year round. With such use it can provide also some moisture to the building which is desirable in regions with winters too dry for comfort.

The range of acceptable temperature in the sun-porch is much larger than within the building proper. Dring day time temperatures up to about 30-35°C are acceptable while at night the porch temperature can be allowed to drop near to the outdoor temperature. But even in this case it serves as an effective wind barrier. Thus the sun-porch not only serves as a buffer zone at night between the building and the outside, but during the sunny hours enables collection and storage of solar energy at higher temperatures than is possible with direct penetration through windows and skylights.

Another advantage of sun-porches over 'solar windows' is the minimization or prevention of damage to fabrics and paints inside the building, as well as the prevention of excessive glare, which usually accompanies direct penetration of sunlight into the indoor space.

On the other hand it should be taken into account that the energy collection efficiency of sun-porches is lower than that of windows. Furthermore, part of the glazing has to be openable or removable to enable ventilation and prevent overheating in summer. The cost of openable glazing which can be closed tightly in winter is higher than the cost of fixed glazing and this fact has to be taken into account in an economic evaluation of sun-porch systems.

Sun-porches can be applied to any type of building, from single storey to high-rise. In the case of single or double storey buildings, the sun-porch can be extended greatly to become an attached greenhouse which can supply the occupants with vegetables and flowers.

c. Southern Walls as Solar Collectors

Southern walls can be utilized in all types of buildings as integrated solar collectors as well as integrated thermal storage elements. There are many architectural and structural solutions for such uses of southern walls so that they can be designed to satisfy different functional, aesthetic and technical requirements.

Solar collecting walls can be either structural or non-structural elements of the building. They can be combined with several thermal storage systems, such as structural storage (within the mass of the building), gravel storage, water storage, etc.

There are some common problems to all types of walls which serve as integrated solar collectors. The first is the prevention of heat load to the building during the summer. As the collection of solar energy is within a wall which may be directly thermally connected to the inhabited space, any absorbed energy causes some heating effect, although the heat may reach the interior with a time lag of several hours.

Because of the annual pattern of the sun in the sky, there is an inherent reduction in the solar load on southern walls in summer because the high elevation of the sun near noon time results in a small angle of incidence of the sun's rays over a southern vertical plane However, in a region with a hot summer, even this reduced heat load caused by heat absorbed in the wall is undesirable.

Many of the integrated solar energy collecting walls utilize thermosyphonic air flow, caused by a temperature difference between the air inside the space behind the glazing, which is heated by the collecting dark surface and the cooler indoor air.

It is convenient to sub-divide the various integrated collecting walls into the following main types:

i. Masonary walls acting as combined collection/storage systems.
ii. Containerized water acting as collection/storage systems.
iii. Insulated lightweight walls acting as solar collectors without heat storage.

These different types will be discussed below.

i. Masonary Walls Acting as Energy Collection/Storage Systems.

This type of a collection/storage wall was developed by Felix Trombe, at the CNRS Laboratory in Odeillo, France and it is usually referred to as the 'Trombe' wall. This approach acquired wide application because of its simplicity and relatively low cost.

In the original design of Trombe, the southern wall was of concrete, 60 cm. thick, and painted in a dark colour on its exterior. Fixed double glazing, 10 cm. away from the concrete helped turn the wall into an integrated energy collection/storage system.

FIGURE 3, B. Hunn house in Los Alamos, N.M. Two storey Trombe wall. Additional direct gain through windows. Supplies 57% of heating requirements.

Solar radiation penetrating through the glazing layers is absorbed in the dark exterior of the concrete wall. The external surface temperature was elevated in the house built by Trombe, up to about 65°C, causing heat flow by conduction into the concrete mass and eventually into the room. Because of the thickness of the concrete wall, a time lag of over 10 hours was achieved between the peak temperature, at the external surface (about 1 p.m.) and the peak of the temperature at the internal surface of the wall.

Openings at the top and bottom of the wall enabled thermo-syphonic air circulation. In winter daytime hours room air entered the space between the concrete and the glazing through openings at the bottom of the wall. The air was heated by contact with the hot black concrete and thus a thermosyphonic air flow was created, with the hot air entering the room through upper openings creating an instantaneous heating of the indoor air.

At night dampers at the openings prevent a reverse thermosyphonic airflow which would increase the heat loss from the building. But even with such dampers, heat is lost from the wall to the outdoors by conduction and convection, because of the relatively low thermal resistance of glazing, even when it is double glazing.

The average solar collection efficiency of the original Trombe wall was about 30%, with about 2/3 of the collected energy being transferred into the building by conduction across the concrete, and about 1/3 of it transferred by the thermosyphonic air flow.

Because of the combination of collection and storage in the same element and its high inertia, this system is effective in providing adequate heating on a cloudy day after a sequence of several sunny days, resulting in extremely high efficiencies. On the other hand, when a clear day comes after a sequence of several cloudy days a significant part of the collected energy is consumed in heating the wall material and is not transferred to the interior, with the result of a low efficiency.

ii. Containerized Water Walls as Combined Collection/ Storage Systems

As an alternative to the Trombe concrete wall, several other systems using containerized water as combined collection and storage were developed. The advantage of water as the thermal storage medium as compared with concrete, is the higher heat capacity, both on the volumetric and the weight basis. Thus, for the same heat capacity, only 40% of the space and 22% of the weight (mass) is required with water relative to concrete. This factor is of special importance in multi-storey buildings, where space is at a premium and weight imposes structural problems and significant extra expense to support the load. Because of the fast heat transfer in water, the whole mass participates instantaneously in the storage space.

An important outcome of this characteristic of water walls is the relatively low temperature of the absorbing surface during daytime, a factor which increases the collection efficiency compared with other collection systems (except direct penetration through windows). On the other hand, at night the external surface temperature of the water wall is relatively high, leading to increased heat loss, unless the wall is insulated at night.

Another characteristic of water walls is the small time lag, as compared with a concrete wall, between the peak of the radiation flux and the heating effect at the interior surface. In those cases where a long time lag is desirable (e.g. about 8-10 hours) this factor should be considered as a disadvantage.

Steve Baer in Albuquerque used water filled oil drums placed behind a glass wall for collection and storage of solar energy for passive space heating. The drums were arranged in racks and painted black on the side facing the sun. Sunlight is admitted directly into the room through the open spaces between the drums and provides some natural light as well as some immediate heating.

FIGURE 4, S. Baer house in Albuquerque, NM. Water storage wall constructed from re-cycled oil drums. Supplies ca. 90% heating requirements

A hinged insulating panel is opened in winter during the day and the solar radiation penetrates through the glass wall and is absorbed in the black faces of the drums. At night it is closed so that the absorbed heat is trapped indoors. In summer, the insulating panel is open during the night so that the water drums can lose heat to the outdoors through the glass wall. During the day, the panel is closed, minimizing heat gain from the outdoors.

Another water collecting/storage wall was developed by Jonathan Hammond of "Living Systems" in Davis, California. In this system, the water is stored in vertical metal pipes, such as culverts. Several houses which apply this system were built in Davis.

Kalwall Corporation is producing translucent plastic tubes (fiberglass reinforced polyester) proposed as water containers in a collecting' storage water wall.

iii. Insulated Southern Walls as Solar Collectors

The Southern walls of all types of buildings can be built as integral solar collecting walls either in a passive or active system. To perform as an integral solar collector, the wall should have high thermal resistance, in order to prevent heat load in summer and to minimize heat loss in winter. From the functional aspects, the interior finish of the wall should conform to all the usual requirements of a wall in terms of appearance, strength, rigidity, etc.

The solar collection function is performed when the external colour of the wall is dark and when the wall is glazed externally. In practice, most commercial solar collectors can be transformed into an integral collecting wall (e.g. like curtain walls) by adding a suitable finishing layer on the interior and by suitable jointing between the panels to render them weather tight.

Such collecting walls can operate in a passive (thermosyphonic) way or actively with pumps or fans transporting the hot medium (water or air). For passive systems, air heating solar collecting walls are more suitable, as the air can flow passively from the energy collecting part of the wall into the adjoining rooms and the heat stored (if so desired) in the mass of the internal building elements such as the ceiling, floor and partitions etc.

The heating effect by a passive lightweight air-type solar collecting wall is almost instantaneous. In practice, a very short time after solar energy impinges on the wall it starts 'producing' a flow of warm air. On a partially cloudy day, such a wall collects and transports solar energy even during short spells of sunshine. In this respect the collecting wall acts in a similar fashion as window collectors. In fact a combination of lightweight insulated walls and southern windows can be very successful for buildings which require sunlight and which are occupied mainly during daytime, such as offices, school-rooms etc.

FIGURE 5, P. Davis house in Albuquerque, NM. Passive air solar collector on sloping ground beneath the house. Rock storage. Supplies 75% of heating requirements.

d. Roofs as Energy Collectors

Several integrated passive and semi-passive systems have been developed mainly in the U.S.A. and Israel, for utilizing the roofs of buildings for collecting solar energy for winter heating and, as we shall see below, night cold for summer cooling.

All these systems have one property in common: they are applicable only to low buildings, usually single-storey, although in some cases the systems can be adapted for two-storey buildings also (e.g. town houses etc.). This limitation is not a serious one in deserts where most of the residential houses and many commercial buildings are likely to have only one or two storeys.

Energy storage in systems using the roof for energy collection can be done either in the roof itself, such as in the case of the Skytherm and Roof Radiation Trap, or in specialized gravel or water storage sub-systems.

Solar collecting panels can be designed and built as prefabricated roof elements. This can be done with both water and air-type solar collectors. The main problem in such an application is effective rain-proofing of the joints between the panels. In addition, when solar collectors are the roof elements, there might be a problem with overheating during summer, when the collectors are not used for space heating. Precautions should be taken to prevent such a heat load on the building in summer.

i. Roof Water Ponds with Movable Insulation.

The 'Skytherm' system has been developed by H. Hay. In this system the whole roof of the building is horizontal and made of steel roofing plates. Above the plates, which receive special treatment against rust, plastic PVC bags filled with water are placed. The lower sheet of the bag is black and the upper is transparent. A second transparent sheet provides some insulation when inflated.

FIGURE 6, H. Hay house in Atascadero, Cal. Plastic water bags on metal roof. Movable horizontal insulating panels. Supplies 100% of heating needs.

Above the water-filled plastic bags there are horizontal movable insulating panels, which insulate or expose the water bags when they cover the roof or are moved aside, respectively. The insulating panels are moved by a motor which is controlled by a differential thermostat according to a pre-set programme by an electronic controller.

In winter, the bags are exposed during sunny hours and insulated at night and on cloudy days. In this way they absorb solar energy when available but their rate of cooling is reduced.

The steel roof is heated by direct contact with the warm bags and the whole ceiling acts as a radiant heating panel. In summer, the system can be used for cooling as will be discussed below.

Another heating and cooling system utilizing water over the roof and operable insulation, has been developed by Jonathan Hammond of Living Systems, in Davis, California. The main difference between the Hammond System and the Skytherm is in the design of the insulating panels. Instead of moving the panels to and fro horizontally, as in the Skytherm system, the "living systems" ponds are covered by hinged lids which are opened and closed by a hydraulic ram. The water in the roof pond is contained in galvanized steel pans with asphalt coating and supported by the roof timber beams, and covered by a transparent plastic vinyl sheet.

The insulating lids are covered on the bottom by a reflective layer. In winter they are hinged open to the south, so that the solar radiation impinging on the lids is reflected towards the water. In this way, the intensity of radiation absorbed in the water ponds is augmented. In addition, in regions with prevalent northern winds in winter, the hinged lids protect the roof ponds from the wind.

ii. The Roof Radiation Trap.

The Roof Radiation Trap has been developed by Givoni in Israel. It is a passive/active system, designed to provide winter heating and summer cooling to buildings with one (or at most two) storeys with a flat roof. The original design was for a concrete roof, but the system can be adapted also to lightweight roofs.

FIGURE 7, Experimental Roof Traps at the Institute for Desert Research, Sede Boqer.

The concrete roof is without attached insulation and the thermal insulation is provided by a layer separated from the concrete roof and sloping up from the northern to the southern ends of the building. The gap between the concrete roof and the insulating layer at the southern side of the building is glazed and provided with a hinged insulating panel, internal or external, which can be closed or opened according to need. Above the insulation, there are corrugated metal sheets painted white externally, to enhance their longwave radiation emissivity. This corrugated metal skin is utilized in summer to collect cold produced by longwave radiation to the sky (see below).

When a long time lag is desired between the peak of solar heating (about 1 p.m.) and the peak of the indoor heating (e.g. in the evening) a gravel layer can be added over the concrete roof, to increase its thermal inertia, time lag and heat capacity, and also to reduce ceiling temperature range.

Under the floor of the building, inside it, or beside it, an insulated gravel thermal storage is provided, where excess heat in winter (and cold in summer, as explained below) can be stored.

In winter, when the Radiation Trap operates in its heating mode, the southern hinged insulating panel is opened during the day and may be closed at night if desirable. Solar energy penetrates during the sunny hours through the glazing at the southern end of the Radiation Trap and is absorbed by the concrete roof, or by the gravel above it.

Part of the heat absorbed in the roof space is transferred by conduction through the roof so that the ceiling acts as a radiant heating panel. The time of the maximum heating depends on the thickness of the concrete roof (and the gravel layer which may be above it). In a residential building, a time lag of about 6-8 hours would be desirable, so that a total thickness of about 30 cm. (concrete roof and gravel) will be required.

The hot air in the space between the concrete roof and the fixed insulating cover layer is blown by a fan to the gravel storage. In this way the roof can act as a short-term thermal storage, e.g. for one night and (with gravel) for one cloudy day, while the specialized insulated gravel storage can serve as reserve for 1-2 additional cloudy days.

III. CLASSIFICATION OF NATURAL COOLING SYSTEMS

Cooling of buildings by natural energies can be effected either by solar energy, although only with 'active' sophisticated systems, or by the utilization of other natural energies which can also be applied in a passive way.

The natural energies which can be utilized for passive cooling are:

a. Nocturnal longwave outgoing radiation
b. Night convection
c. Water evaporation

Combinations of these cooling sources are also possible, such as combining convective and evaporative nocturnal cooling or combining nocturnal longwave radiant initial cooling of air with subsequent super-cooling or water evaporation.

The applicability of the various natural cooling sources (besides solar energy) depends greatly upon the climatic conditions prevailing in summer at night in a given region, and in particular on the dry and wet bulb temperatures of the nights in summer. This is because the low level to which a thermal storage can be cooled is close to the dry or wet bulb temperaatures at night, for convective or evaporative cooling, respectively.

This section discusses the applicability of various cooling systems utilizing the above mentioned natural energies and some of the problems involved. Quantitative comments are related to the climatic conditions prevailing in various deserts in the U.S.A. and the Middle East.

a. Cooling by Nocturnal Radiation

Any element of the external envelope of the building which 'sees' the sky loses heat by the emission of longwave radiation (with peak radiation at a wavelength of about 10 microns). As the roof is the building element most exposed to the sky it is the most effective longwave radiator.

Some of the gases in the atmosphere, especially water vapour as well as CO_2 and dust, absorb and emit longwave radiation. Thus at night there is a balance between the radiation emitted by the roof towards the sky and the downward radiation from the atmosphere. Only the net radiant heat loss is effective in cooling the building.

The net radiant exchange depends on the temperature difference between the emitting surface which is following the ambient air, and the 'effective' temperature of the atmosphere. This difference increases with the rise in the ambient air temperature, as the temperature of the atmosphere is more constant than the air temperature near the ground.

Clouds are very effective in 'blocking' the outgoing radiation and therefore the net radiant loss decreases greatly with cloudiness.

The following table can serve to estimate the net longwave radiation heat loss from a horizontal surface under a clear sky and different conditions of air temperature and vapour pressure.[4]

Net longwave radiant loss (kcal/hr·m^2)							
Vapour pressure (mmHg) / air temperature	4	6	8	10	15	20	30
20°C	110	100	90	80	70	-	-
30°C	130	115	105	95	80	75	70

When the sky is cloudy the outgoing radiation decreases. Geiger[5] gives the following numbers as percentages of the values for cloudless sky.

Cloudiness in tenths	% outgoing radiation
0	100
1	98
2	95
3	90
4	85
5	79
6	73
7	64
8	52
9	35
10	15

When the emitting surface is at a higher temperature than the ambient air, it can utilize all the potential of radiant heat loss. However, when the emitting surface is initially at the same temperature as the ambient air, the emittance of longwave radiation lowers its temperature below the ambient air level. In this case a convective heat gain from the air counter-acts the radiant heat loss till an equilibrium is reached between the heat gain and loss. The convective heat gain depends on the wind speed near the emitting surface so that as the wind speed increases the temperature of the emitting surface is close to that of the ambient air.

Under ordinary conditions, the cooling effect resulting from the net radiant heat loss by the roof cannot be used directly for cooling the building, because of the thermal resistance of the roof. Almost all the "cold" produced is wasted to cool the ambient air in contact with the roof unless the radiating surface is at a higher temperature than the ambient air.

In order to utilize the cooling effect of the nocturnal radiation, ways have to be found by which the cold produced is transferred into the building's interior. Each of the Hay "Skytherm," Hammond "Cool Pool" and Givoni "Radiation Trap" systems discussed above incorporate features to enable cooling by nocturnal radiation.

i. Cooling by the Skytherm system

In the Hay Skytherm system, plastic water bags over a metal roof are exposed in summer at night to the sky. The plastic material is partially transparent to longwave radiation, the water emits radiation to the sky and is cooled by this process. In addition, as the water temperature at night is usually above the ambient air temperature, because of the heat absorbed in it during the day, there is also heat loss by convection from the water bags to the ambient air.

The water bags are in contact with the metal roof and thus the ceiling acts in summer as a cooling panel for the rooms under it. During the daytime, the ceiling absorbs heat from the rooms underneath, which penetrates through the windows and the walls. The water bags are insulated during daytime hours by movable insulating panels above them so that heat gain from the outdoors is greatly reduced.

Performance of the Skytherm as a cooling system[6]

On a typical summer day in Atascadero, when the outdoor temperature range was from 13 to 35°C, the bag water temperature range was 18.5 - 20°C and the indoor air temperature range 21-22.5°C. Thus the indoor range was less than 10% of the outdoor range and the average indoor temperature was about 2°C below the outdoor average. Most of the night, the water temperature was appreciably above the ambient air so that convective heat loss was a significant factor in the cooling process of the water bags.

ii. 'Living Systems' Roof Pond

As with the Skytherm system, here too the pattern of opening and closing the lids is reversed in summer. The water ponds are exposed during the night and insulated during the day. The bottom of the water pans is directly exposed to the indoors so that the pans act as convective and radiative cooling elements

Performance[7] of the 'Living Systems' Roof Ponds

A house was built in Davis utilizing the 'Living Systems" Roof Ponds. In a typical summer week, with outdoor temperatures ranging from about 15.5°C to about 43°C the indoor temperatures ranged from about 21°C to 26.5°. Thus it has been demonstrated that this system can provide adequate cooling in the hot summer of the Central Valley of California.

iii. Cooling by the Roof Radiation Trap System

In the Givoni Roof Radiation Trap system, the insulation layer is covered by corrugated metal sheets, painted white on their external side. Painting is necessary because of the low emissivity of metals, so that the paint layer serves as the radiating surface.

During daytime hours, the system is not activated at all and the roof's metal temperature is established by equilibrium between the absorbed solar radiation (Minimized by the white paint) and the heat loss by longwave radiation to the sky and by convection.

At night the painted metal roof is cooled by the emitted longwave radiation to a temperature below the ambient air. The temperature drop, under passive equilibrium, depends on the cloudiness of the sky, ambient vapour pressure and the wind conditions. Typical values are from 4 to 9°C below ambient air.

To utilize the cold produced by the longwave radiation, air is sucked by a fan under the corrugations of the metal roof. The air is cooled, by contact with the cold metal, to a temperature 1-3°C below the ambient air, depending on the flow rate and the environmental conditions.

The cooled air is passed through a gravel storage and thus the cold is stored to be utilized in cooling the building during the following day.

In experiments at the Institute for Desert Research in Israel,[8] a gravel mass of about 0.2 m^3/m^2 of roof area was cooled to a temperature approximately equal to the ambient air minimum temperature. During daytime hours when cooling is required, indoor air is circulated through the cold gravel and is cooled to a temperature close to the gravel temperature. In this way, the gravel absorbs the heat penetrating into the interior of the building across its envelope and by infiltration.

b. Cooling by Night Convection (Natural and Mechanical)

In warm regions where the night temperature in summer is below the comfort range (e.g. below 20°C) it is possible to store the coolness of the night air, either in the structural mass of the building and/or in a specialized gravel storage. The air flow at night through the building or through the gravel mass can be induced naturally by the wind (where wind speed at night is sufficient, e.g. above 3 km/hr) or mechanically by a fan. During the following day, it is possible to utilize the stored cold so as to keep the indoor temperatures within the comfort range provided that the building is not ventilated by the outdoor hot air during the daytime hours.

In practice it is possible to apply this approach in regions where the ambient vapour-pressure in summer is below 15 mmHg, because only then the human body can feel comfortable without feeling air motion at temperatures up to about 26-28°C, according to the humidity level. In Israel, such conditions exist throughout the desert.

Furthermore, to insure success by this approach, the building should be of a high thermal quality (G coefficient less than 1.5) and have effective protection in summer from solar heat, by proper shading of the windows and a reflective colour for the external opaque envelope. The mass of the building should be insulated at the external side of the main mass of the external walls and the roof.

When the storage of the night-cold is in a gravel mass there is plenty of surface area for heat transfer between the air and the storing mass. However, when thermal storage is contemplated in the structural mass of the building, the surface area for heat transfer may be a major factor in determining the actual heat transfer between the storing mass (walls, floor, ceiling, partitions, etc.,) and the air flowing through the building.

In estimating the potential for cooling by storing night cold in a gravel mass it should be taken into account that the average gravel temperature can be brought at the end of the night to a temperature about 1-2°C above the ambient air minimum. The upper temperature limit for the use of the gravel for cooling the interior is about 2°C below the upper limit of the comfort zone. Thus the upper temperature limit for the gravel depends on the outdoor humidity level and may range from about 23°C in humid regions (vapour pressure of about 20 mmHg) to about 25°C in regions with dry summers (Vapour pressure less than 15 mmHg).

When night coolness is stored in the structural mass it is not possible in practice to cool the mass at night to the same temperature attainable with gravel cooling. The minimum temperature of the structural mass, even with good nocturnal ventilation, can be brought to about 3°C above the outdoor minimum. The upper temperature limit for indoor cooling (from the comfort viewpoint) is about the same as that of the gravel. Therefore, the useful range of structural thermal storage is smaller (by about 2°C) than the range of the gravel storage.

c. Night Evaporative Cooling

Water evaporation is used extensively in arid regions to cool the indoor air of buildings (e.g. by desert coolers). However, such systems have to be operated mainly during the day, when the cooling is mostly needed, and consequently have to cool large quantities of very hot air (in the range of 35-42°C). From the physiological viewpoint this approach is not always desirable for two reasons:

i. The cooled indoor air is excessively humid.
ii. The high rate of air flow and air change which are necessary for effective cooling, cause very great variation in the air speed, and the associated thermal sensation, within the cooled building, in addition to the waste of energy to cool the air which is subsequently discharged to the outside.

Another possible approach to evaporative cooling is to obtain it during the night hours, sometimes in conjunction with longwave radiant cooling and convection, and to store the cold in a thermal storage for daytime or continuous use.

It is possible to compare the expected performance of daytime evaporative cooling with that of night cooling stored for daytime utilization by considering the different combinations of air temperature, vapour pressure and wet bulb temperature, as shown on the psychrometric chart.

For example, the conditions typical of the Negev Region in Israel will be considered. The assumed average daytime conditions are: air temperature 32°C, vapour pressure - 13 mmHg and W.B.T. - 20°C. The assumed average night conditions are: air temperature - 21°C, vapour pressure - 10 mmHg and W.B.T. - 14°C.

With evaporative cooling, the air temperature is reduced and the vapour pressure is elevated along with the W.B.T. In practice saturation of the air is not usually achieved, so that an 80% approach has been assumed.

With daytime evaporative cooling, the air is cooled to about 22°C with a vapour pressure of 16 mmHg, while with night evaporative cooling the temperature of the air is 16.5°C and the vapour pressure is 12 mmHg.

The amount of water needed to produce these cooling effects is much greater for daytime than for night cooling. Thus, every m^3 of air which is cooled during the day requires the evaporation of about 4 grams of water, while the night evaporation consumes only about 2 grams.

Evaporatively cooled night air can be utilized to cool a gravel storage bed by blowing the cooled air through it. In practice the water can be sprayed directly over the gravel, which usually is at a temperature higher than the dew point temperature of the night air. In such a case evaporative cooling can be obtained even when the outdoor air is at 100% relative humidity, because of the higher vapour pressure of the water on the warmer gravel surface.

In humid regions, night air which is super cooled by contact with a surface cooled by longwave radiation, can often reach a temperature below the ambient dew point, and precipitate some of its water content by condensation over the radiating surface. The air thus becomes drier and can be used more effectively for combined evaporative and convective cooling of the storage gravel bed.

An alternative approach, effective where high rates of air change are required (e.g. schools) is the so-called rock-bed generator. The evaporative cooling of the exhaust air chills a switched-bed rock-filled recuperator. Incoming air is then cooled by passing through the previously chilled rocks, only a small amount of moisture being added. The air flow is switched every ten minutes. This system requires a conveniently designed storage unit, but the required storage area is not large.

IV. REFERENCES

1. Proceedings of the Passive Solar Heating and Cooling Symposium, Albuquerque. (1976) Los Alamos Scientific Laboratory

2. B. Anderson, The Solar House Book. (Cheshire Books, Harrisville, New Hampshire, 1976).

3. D. Watson, Designing and Building a Solar House. (Garden Way Publishing. Charlotte, Vermont, 1977).

4. B. Givoni, Man, Climate and Architecture, 2nd ed. (Applied Science Publishers, London, 1976).

5. R. Geiger, The Climate Near the Ground. (Harvard University Press, Cambridge, Mass. 1959).

6. H.R. Hay, in The Sun in the Service of Mankind. (UNESCO, Paris, 1973).

7. David A. Bainbridge, "Natural Cooling: Practical Use of Climate Resources for Space Conditioning in California" Proceedings, 3rd Annual Solar Cooling Conference, San Francisco,(1978)

8. B. Givoni, Energy and Buildings. 1. 141 (1977)

Some Details About the Authors

L. Berkofsky: Professor. Head of the Desert Meteorology Unit at the Blaustein Institute for Desert Research, Sede Boqer. Joint appointment with the Dept. of Geography, Ben-Gurion University of the Negev.

D. Cohen: Professor. Director of the Isan Center for Comparative Medicine in Tel Sheva. Head of Unit for Animal Physiology in Arid Zones at the Blaustein Institute for Desert Research, Sede Boqer, Ben-Gurion University of the Negev.

M. Evenari: Professor Emeritus, Hebrew University of Jerusalem. Head of the Run-Off Farms Unit at the Blaustein Institute for Desert Research, Sede Boqer, Ben-Gurion University of the Negev.

D. Faiman: Associate Professor. Head of the Applied Solar Calculations Unit at the Blaustein Institute for Desert Research, Sede Boqer. Joint appointment with the Dept. of Physics, Ben-Gurion University of the Negev.

J. Gale: Associate Professor, Hebrew University, Jerusalem. Head of the Closed System Agriculture Unit at the Blaustein Institute for Desert Research, Sede Boqer, Ben-Gurion University of the Negev.

B. Givoni: Professor. Head of the Unit of Solar Buildings and Energy Conservation at the Blaustein Institute for Desert Research, Sede Boqer, Ben-Gurion University of the Negev. Joint appointment with University of California at Los Angeles.

Y. Gradus: Senior Lecturer. Head of the Dept. of Geography, Ben-Gurion University of the Negev. Head of the Geostatistical Unit at the Blaustein Institute for Desert Research, Sede Boqer.

A. Issar: Associate Professor. Holder of the Alain Poher Chair of Desert Research. Head of the Hydrology and Water Resources Engineering Unit at the Blaustein Institute for Desert Research, Sede Boqer, Ben-Gurion University of the Negev.

E. Marx: Professor, Tel Aviv University. Head of the Nomad Settlement Unit at the Blaustein Institute for Desert Research, Sede Boqer, Ben-Gurion University of the Negev.

A. Rahamimoff: Senior Lecturer. Head of the Desert Architecture Unit at the Blaustein Institute for Desert Research, Sede Boqer, Ben-Gurion University of the Negev.

U. Regev: Senior Lecturer, Dept. of Economics, Ben-Gurion University of the Negev. Head of the Resource Economics Unit at the Blaustein Institute for Desert Research, Sede Boqer.

A. Richmond: Professor. Director of the Blaustein Institute for Desert Research, Sede Boqer, and head of its Hydrobiology Unit. Joint appointment with the Dept. of Biology, Ben-Gurion University of the Negev.

M. Shachak: Lecturer. Head of the Enviornmental Education Unit at the Blaustein Institute for Desert Research, Sede Boqer. Joint appointment with the Dept. of Biology, Ben-Gurion University of the Negev.

E. Stern: Lecturer, Dept. of Geography, Ben-Gurion University of the Negev. Joint appointment with the Geostatistical Unit at the Blaustein Institute for Desert Research, Sede Boqer.

A. Weingrod: Professor. Chairman of the Dept. of Behavioural Sciences, Ben-Gurion University of the Negev. Head of the Social Studies Unit at the Blaustein Institute for Desert Research, Sede Boqer.

M. Zohary: Professor Emeritus, Hebrew University, Jerusalem. Member, National Academy of Science, Israel. Head of Applied Geobotany Unit at the Blaustein Institute for Desert Research, Sede Boqer, Ben-Gurion University of the Negev.

AUTHOR INDEX